The Life of a Forestry Ploughman
and other adventures

The Life of a Forestry Ploughman
and other adventures

Peter Weir

MÒR MEDIA LIMITED

The Life of a Forestry Ploughman
and other adventures

Copyright © 2020 Peter Weir

All rights reserved.

All rights reserved. No part of this publication may be reproduced or distributed in any form or by any means, or stored in a database or retrieval system, without the prior written permission of the author.

Peter Weir asserts his moral right to be identified as the author of this book.

Photographs ©Peter Weir and ©Paulette Weir, unless otherwise stated

ISBN 978-1-9993668-3-4

First published 2020

Mòr Media Limited
Argyll, Scotland

www.mormedia.co.uk

Design by Helen Crossan

CONTENTS

Photographs and Map	vii
Acknowledgements	ix
A Brief History of Forestry in Scotland	1
My Life So Far	5
Places Where I Worked	11
Strathlachlan	13
London	21
Kilberry	23
Lochgoilhead	25
Fearnoch and Taynuilt	27
Glen Creran	29
Appin	35
Ballachulish and Glen Coe	37
Rannoch Moor	53
Barcaldine and Bonawe	57
Tyndrum	59
Crianlarich	63
Dunoon and Arran	67
Tighnabruaich and Kilfinan	69
Tarbet, Loch Lomond	73

Arrochar	75
Glen Etive	77
Bridge of Orchy	83
Dalmally	87
Glen Orchy	93
Glendaruel	97
Inverkip	101
After the Forestry	105
Our Restaurant Time	113
Conclusion	115
Kilmartin	117
In Closing	119
About the Author	121

PHOTOGRAPHS AND MAP

1.	Near Dalavich	3
2.	Old Castle Lachlan, near New Castle Lachlan at Strathlachlan	4
3.	The old phone box where we stopped every day	4
4.	Strathlachlan Primary School	6
5.	The bridge below the school	6
6.	Strachur Shinty Team, circa 1900	8
7.	Strachur Shinty Team, in 1936	8
8.	Some of Strachur Shinty Team, circa 1960	9
9.	Old Castle Lachlan	12
10.	Castle Lachlan, Strathlachlan	12
11.	Working on the Tighnabruaich Road	16
12.	Kyles of Bute from Tighnabruaich Road Viewpoint	17
13.	Newton Village from where we set off in the boat	20
14.	Forestry ploughing	22
15.	Deep draining plough attached to two tracked tractors	22
16.	The Paps of Jura	23
17.	Forest at Lochgoilhead	25
18.	Ben Cruachan from Fearnoch	27
19.	Glen Creran where my friend's grandfather's house was	29
20.	Michael Palin, son Tom and Ang Phurba, Aonach Eagach Ridge	32
21.	Michael Palin, son Tom, me, and Ang Phurba	32
22.	Castle Stalker, Appin (Castle Aaargh)	33
23.	Michael Palin, son Tom and Ang Phurba, Aonach Eagach Ridge	34
24.	Robin's Farm in Appin looks out towards Castle Stalker	35
25.	Hamish MacInnes filming a climb for the BBC	37
26.	Old Inverigan House in Glen Coe	44
27.	The old farm ruins at Brecklet	45
28.	One of the ruins at Brecklet	46
29.	The Glencoe Mountain Rescue Team in the 1980s	47

30.	Walter Elliot and sons, Walter and Willie, about 1946	47
31.	John Smith and myself, the author	49
32.	John Smith and friends, with cameraman John Boyce	49
33.	Letter from John Smith	50
34.	Glencoe Ski Centre	50
35.	Paul Moores and myself at the Glencoe Ski Centre	51
36.	Rannoch Moor	53
37.	On the road to Bonawe	57
38.	Approaching Tyndrum from the A85	59
39.	Looking towards Ben More, Crianlarich	63
40.	Hamish MacInnes	64
41.	A film Hamish MacInnes was making about the Glen Coe area	66
42.	Tighnabruaich and Kilfinan area	69
43.	Tarbet, Loch Lomond	73
44.	Loch Long looking towards Succoth	75
45.	Glen Etive looking towards Beinn Fhionnlaidh	77
46.	Ski Lodge, Bridge of Orchy	83
47.	Glen Orchy road end looking towards Dalmally	87
48.	Chris Bonington and myself	91
49.	Glen Orchy (from the bridge)	93
50.	Glen Orchy, where we used to cross the river in the Land Rover	94
51.	Looking to Glendaruel from the Tighnabruaich Road Viewpoint	97
52.	Looking to Inverkip Forest from Dunoon	101
53.	Mural of my wife's ancestors in Old Albuquerque, New Mexico	108
54.	1933 Newspaper article featuring interview with Mrs Dow	109
55.	Billy the Kid's grave in Fort Summer, New Mexico	110
56.	Graves of Billy the Kid and his companions	110
57.	Some ancestral places, marked	116
58.	Standing Stones at Kilmartin	117
59.	The foot indentation where the Pictish kings were crowned	117
60.	Glen Etive area	119
61.	Peter and Paulette	120

ACKNOWLEDGEMENTS

My grateful thanks to my wife, Paulette, without whom this book would not be possible. She has devoted a considerable amount of her time to managing this project, including correspondence, photography, and the initial editing of the text. Thank you, Paulette.

A BRIEF HISTORY OF FORESTRY IN SCOTLAND

Most of Scotland was covered in native forest for thousands of years after the ice age. The native forests were mostly hardwoods, as well as pine, alder, rowan, and birch. Years ago, the climate changed, and a period of wet weather began which was not good for the trees to grow, especially the Caledonian pine trees whose roots are still visible in peat bogs on Rannoch Moor and other areas. Many species of wildlife lived in these forests: bears, wolves, wildcats, lynx, wild boar, deer, foxes, badgers and birds. There were no wire fences, and certainly no deer fences.

Then along came humans. The last wolf was reportedly killed in Scotland in the late 1600s. Trees were felled for use in heating and building homes and other structures, as well as to allow for agriculture, sheep farming and sporting estates. Today, Scotland has very little remaining native forest.

The majority of forests in Scotland are now commercial plantations. Most planted forests in Argyll and the Highlands take about fifty to sixty years to mature, as the trees are mostly Sitka spruce, one of the fastest growing species in Scotland's wet climate. In these plantations, which are often surrounded by deer fencing, very little wildlife exists. When the Forestry Commission bought land from the landowners, they erected deer fencing around the whole area to keep animals out, and any deer remaining trapped within the deer fencing were shot so that they would not damage the young trees. Once the trees mature, the dense growth does not allow light in, therefore little or nothing grows below the trees. Most animals and most birds are not able to live in the dense forests. There may be mice and some badgers or foxes. When animals are removed from an area because of human intervention, the birds of prey also disappear.

There is a big demand for milled wood and caber board for construction. Pulp is used for paper and other products; in some countries, pulp is also used for making rayon and viscose for clothing. The Forestry Commission provides the timber for this. Once the trees have been felled and removed, the area is replanted and the cycle continues.

World War I began in 1914 and trees were felled and used for the war effort. Following the end of the war in 1918, the Forestry Commission was formed in 1919 to improve Britain's timber supply. World War II began in 1939 and again timber was needed for the war effort. The Forestry Commission gave people living in remote areas full time work; however, the wages were low.

This book takes you to all the places I have worked, with a few adventures added in here and there as I remember them.

I was employed by the Forestry Commission from the late 1960s for about twenty years. My father also worked for the Forestry Commission. It gave us work which we appreciated. My wages were low; however, I was given subsistence for working away from home. I was employed to plough up the hills for planting trees with a tracked tractor. I also deep drained newly-ploughed ground and areas that had been planted. Deep draining involved two tracked tractors with wide tracks and about a three-metre towrope in between, as well as a big plough with two rear tractor tyres on each side which, when working right, would be used to dig a drain about a metre deep. The length of the whole deep draining system was about thirty metres. When turning on the mountain it caused a bit of destruction, especially when the trees were about ten feet high or more. When you went through big trees, they were falling and going under the plough. They became jammed up under the plough, and as there was nowhere to put them, we came out of the forest with the plough sitting four to five feet high off the ground. Old Caledonian pine roots also caused problems when deep draining. The plough would not break them out of the ground, and we would keep breaking the shear pin (shear pins were made of softer metal to stop damaging the plough), so deep draining did not work in these areas. It worked quite well where the trees were either very small or the area had not been planted.

When I started ploughing, the hill ploughing involved a tracked tractor with a trailing plough behind. When working on steep ground, the trailing plough proved to be dangerous when turning on a steep slope. After a few incidents, the Forestry Commission decided to fit the plough onto the back of the tractor so the driver could reverse up the mountain until it got too steep and then the driver would drop the plough which would stop it from sliding down the mountain. Most of the ground we ploughed was hill ground and had never been ploughed before, except where there were old settlements. We were told not to touch the ruins. We ploughed and deep drained twelve months a year, except for holidays, deep snow and breakdowns. We also ploughed when birds were nesting. We could not see the nests for the rough terrain; the only thing we saw were feathers or an odd bird flying out below our tracts. We ploughed in rare plants, rare mosses, rare trees, rare shrubs, rare insects, nests of adders, nests of stoats and weasels, and changed water courses. Scottish Natural Heritage (SNH) was not formed until 1992. Prior to that there was the Nature Conservancy Council, formed in 1973 to replace the Nature Conservancy, founded by Royal Charter in 1949.

It is easy to look at the land—the hills and mountains—and imagine all the people who lived in the old ruins before the Highland Clearances and what it would have been like without the ploughed-up, closely-grown forests, and loss of flora, wildlife, sheep and habitat.

1. Near Dalavich
Many areas were like this before deep draining and hill ploughing

2. Old Castle Lachlan in the background, near New Castle Lachlan at Strathlachlan

3. The old phone box where we stopped every day to phone the lady in the exchange
The phone box is over sixty years old.

MY LIFE SO FAR
(AS I CAN VAGUELY REMEMBER IT…)

I was born in the gatehouse at Castle Lachlan, Strathlachlan. My father worked at the castle. When I was a baby in nappies, a goose or goosander picked me up by the nappy and carried me into the wood. Apparently, my father came to my rescue. The joke back then was the goose took me for a gander.

There were about seven to ten pupils at school. I can't remember learning much about anything, as I spent most of the time playing up in the hills. The school toilets were about fifty yards below the school, on each side of a bridge which went across the river. They consisted of two tin huts with wooden seats and a hole in the middle. Being boys, our fun was to go down below the bridge along a wee ledge and wait for one of the girls to go to the toilet, fill a small tub up with water and throw the water up the hole into the girl's toilet.

One of the other things that stands out in my mind is that we used to go across the field (about one mile) to raid the farmer's orchard. One day, the farmer caught me. He proceeded to take his belt off his trousers, then pulled down my shorts, put me across his knees and belted my bare bottom. We never went back to raid his orchard.

SCHOOL

The school was in a very quiet area with a single-track road, which was the old Tighnabruaich road, up about a hundred yards from the school. The highlight of our day was, if we heard a car coming in the distance, to run up to the side of the road and wave to the car. As there were not many cars on the road at that time, the regular car drivers would sometime stop and give us sweeties. We had to walk to school in my early days, and I remember that on the way home we would stop at a red telephone box. Back then you would pick up the phone and phone the exchange to get your number connected. We got to know the lady in the exchange quite well, as we phoned her nearly every day just to say hello and she always had a wee chat with us.

This was just after the Second World War and, with not a lot of money to feed us, life was quite tough for our parents. When we came home from school, we would sometimes have sugar pieces: bread coated in sugar.

When I was in primary school, I discovered skiing. I remember going into my father's shed and getting two flat bits of wood, about two yards in length. I nailed a block of wood onto the front of each one and then shaped it to

4. Strathlachlan Primary School
(one of my ancestors who was born in 1756 was a teacher in Strathlachlan school—probably in an older building)

5. The bridge below the school.
The girls' and boys' toilets were on either side of the bridge.

form a curve. I then proceeded to get an old pair of sandshoes and nailed one in the middle of each length of wood. There had just been a snowfall, so off I went with the new skis over my shoulder. I headed for the field just above our house. The plan was to take off my boots and tie them around my waist, put my feet in the sandshoes fitted to the skis and off I would go. After a few attempts I managed to stay upright. However, something I forgot to do was to learn to stop! Thank goodness for the fence at the bottom of the field or I would have ended up in the stream.

One day I decided to go to another field, as the snow cover was a bit better. When I looked down the field, I saw another person coming up the field carrying skis. It was a lady I knew from a small village down below the road and she looked as if she knew how to ski. It was Sheila, a retired schoolteacher (not mine), and when she saw my homemade skis, she had a good laugh and said, 'Peter, I have another old pair of skis down at the house and you can have them.' The snow lasted a few days and Sheila taught me the basics of how to ski—especially how to stop. Many years later, I took up skiing on actual ski slopes and became the manager of the Glencoe Ski Centre for about twelve years.

SHINTY

My grandfather (Duncan), my father (Duncan), and I all played shinty.

6. Strachur Shinty Team, circa 1900
My grandfather, Duncan Weir, is in the back row, third from left

7. Strachur Shinty Team, in 1936
My father, Duncan Weir is in the back row, second from left

8. Some of Strachur Shinty Team, circa 1960
Yes, that is me, Peter Weir, first on the left

A TRIP TO GLASGOW

When I was young and living in the forestry house in a small village called Leanach, one of the owners of Rangers Football Club, Mr John Lawrence, bought a holiday home in a wee village called Leachd by the side of Loch Fyne, just down the road from where I lived. I can't remember how it came about, but he decided to take all the children in the area to Rangers' Football Stadium at Ibrox in Glasgow. As there were not too many children in the village, he hired a minibus, which he drove himself, and took us all to Ibrox. He treated us to food and soft drinks on the road on our way to Ibrox, and on the way back. When we arrived there, as it was in the middle of the week, there were no football games being played, so we could go anywhere in Ibrox, which we did, with Mr Lawrence being our guide. We had the most fantastic time looking around the trophy room, the football pitch, the changing rooms, and anywhere we wanted to go. What a wonderful man, and what an experience for us small children living out in the country.

PLACES WHERE I WORKED

9. Old Castle Lachlan
(where I had the encounter with the ghost [sheep])

10. Castle Lachlan, Strathlachlan
I first started work here, after leaving school

Strathlachlan

I worked at the castle farm after leaving school. One day, I was down at the old Castle Lachlan with the tractor and trailer feeding the cattle when a big thunder and lightning storm started. Back then there were no cabs on the tractors. It was lashing down with rain, lighting and thunder, so I headed into the old castle for shelter. Inside the doorway was a big open space and to the right there were dungeon-type rooms with no light. I ran into a room (dungeon) to seek shelter. With the lighting and thunder all around, I looked over into the corner of the room and all I saw were two eyes staring at me. Being only fifteen years old, I got such a fright I ran out of the old castle, followed by—a sheep!

When I was living in Strathlachlan, my father was working at Castle Lachlan. My father told me a great story about a pony in the field by the old Castle Lachlan. My friend, Ewan, who is now chief of the Clan MacLachlan and lives at the new castle, had a pet pony in a field down at the old castle and one night there was a thunderstorm and the pony went missing. After a day or so searching for the pony, it was reported to have been seen across Loch Fyne on the other side at Minard. Loch Fyne at that point is about one and a half to two miles wide and to get to it by road would be about eighty miles. The pony swam the loch. The MacLachlans went to Minard by road to bring the pony back.

There was a local story that, at a certain time of the year, you would see the ghost of a white horse riding around the castle. The chief of the Clan MacLachlan was fighting with Bonnie Prince Charlie at Culloden and was supposed to have been the aide-de-camp to him. He was killed at Culloden and his white horse is supposed to have returned to the old castle by swimming across the loch. I saw a documentary with Paul Murton on television which mentioned a Lachlan McLachlan who was ordered by Bonnie Prince Charlie to ride in front of the Scots army on a white horse and get them to charge forward. No sooner had he mounted his horse than a cannon ball took his head off. This was what supposedly caused the failure at Culloden. After Culloden their estate was forfeited, and the castle was garrisoned by the Campbells. One day, an English warship sailed up Loch Fyne and destroyed the castle. Eventually the estates were returned to the MacLachlans. We spent many a night at the old castle in my teenage years, having a few drinks and looking for the ghost of the white horse—until the spirit ran out.

Another story I remember from my days at the castle was when I was working on the shore with a digger, and loading a lorry with gravel; this was common back then. A man appeared carrying a jerrycan (fuel can) and approached me and asked if I had any diesel, as he had run out of fuel in his boat. I replied that I would have to go up to the farmyard to get some when I finish loading the lorry. Off I went with the man in the digger, filled the can up and returned to the shore. As the lorry had not returned, I took the man along the track that went along the shore for about one mile to where it stopped.

I asked him where his boat was, and he replied just over the hill. The man got off the digger with his fuel can and I returned to load the lorry which had not arrived. When the lorry arrived, I loaded it, but something in my mind made me go back along the shore to check on the man, as I had not seen a boat going down the loch.

There was no sign of him, so I headed over the hill to a bay with a small island, and still no sign of him, so I went over another the hill to another bay that had a small rockface. I spotted the boat close to the rockface but no sign of the man. As I approached the boat, I spotted the man, trapped between the boat and the rockface, and holding onto the boat. I jumped on board and pulled the man aboard. He was in a bad way and could not speak. He had managed to get the can of fuel on board but had fallen into the water and could not swim. I left him on deck to recover and continued to put fuel in his tank. He recovered and when we got the boat started, he said 'thank you', and off he went. I never saw him again.

The kitchen at the castle was the most important place when you were young and always starving. The cook was called Maggie Blue. She always made sure we had plenty to eat and would also tell us scary tales about the castle and the surroundings. She lived at the gate lodge where I was born, as our parents had moved to a forestry house about four miles away. One of Maggie Blue's ghost stories was about the old walled orchard which I had to cycle past every day going to work. Most of the road on each side was covered in rhododendrons and trees—quite spooky when you were young. She also had to walk past the orchard to get to her work. She would tell the story of how, walking home in the dark she would often hear the sound of children singing and laughing coming from inside the walled orchard. One morning, when I was about fifteen years old, I was cycling to work in the dark. Back then, I had a torchlight that clipped on to the front of the bike. I was passing the old orchard on the gravel road and thinking about Maggie Blue's ghost story when—all of a sudden—the torch jumped off the bike and the light went out. I got such a fright that I threw the bike and broken torch into the bushes and ran the rest of the way to the castle.

Another story from my days at the castle was when my pal Ewan and I were cutting silage in a field. Ewan was cutting the grass and I was running along beside him on another tractor and trailer, collecting the cut grass. For some reason or other we were shouting to each other and an argument started. We both jumped off the tractors and had a punch-up in the field. The tractors kept going (with no drivers), cutting a trail all over the field until we made up, ran after the tractors, jumped on and stopped them. We are still friends to this day.

When we were children, we lived close to where there was a training area during the Second World War. It was very common to find .303 cartridges in the old stone walls on the hill. We used to remove the bullet from the cartridge and make bombs for rabbits, as food was a bit short back then. I will not go into detail how we made the bombs. In the field where there were lots of rabbits, we used to block up as many rabbit burrows as possible and leave two open with a net over one and the other for the bomb. On one occasion, the bomb went off, there was smoke coming out all over the field, and we never caught a rabbit. The rabbits were out in the field next day!

Around this time an old man told me how to catch a pheasant, as they were quite common back then. He told me to get a paper bag, put some corn in it, then narrow the neck of the bag, put sticky tape around the inside of the neck of the bag and leave a trail of corn up to the bag. The pheasant would eat its way up, stick its head inside the bag and the bag would get stuck around its neck. I never caught a pheasant. Thank goodness for the ration book that was issued to every family back then.

After the farm, I went into the Forestry for a very short time, brashing trees—cutting the lower branches off to give walking access. Because he did not trust us, the forester in charge used to tie a tape onto the tree after he measured our work, and would peel an orange and leave it on the ground beside the tape, so we just moved the tape and the orange peel to suit us, as the pay was very poor.

I worked with some real characters back then, but one man stands out. He was an old man near retirement, and every morning he would turn up for work wearing his pyjamas under his work clothes. When the old man died, his brothers took him and the coffin in the work van to the cemetery, with the coffin sticking out the back of the van.

I worked on the Glendaruel and Tighnabruaich roads, driving excavators for about two years. The new Tighnabruaich road was a new road through very rough but beautiful mountain terrain. Back then, we did not have a cab on some of the excavators, as you can see in the photo. We worked twelve hours a day, and a half-day on Saturday. There were no tea huts in most areas, and so for tea and lunch breaks I would turn the bucket upside down and have my breaks under the bucket as it was often raining.

11. Working on the Tighnabruaich Road (me, aged eighteen)
This photograph featured on the cover of a calendar for International Harvester

The Life of a Forestry Ploughman and Other Adventures

The reason I got to drive such a big and brand-new machine was because all the men who had worked with the firm for years were arguing about who was going to get the new machine. The boss (also the owner of the firm) appeared one day where I was working on a small machine and said, 'Come with me.' In his Land Rover, we drove up a rough track which is now the Tighnabruaich Road, and there at the new viewpoint was the brand-new machine.

The boss said, 'Here is your new machine.'

I said to the boss, 'I cannot drive that big machine.'

He replied, 'You will soon learn,' and I did—there was a massive vertical drop where I started work on the machine.

When I was working on the Tighnabruaich and Glendaruel road and the machine had broken down, I would sometimes help with the drilling of rock and the blasting. Back then the explosives were easy to come by.

One day, my friend and I decided to blow up a pool in our local river, so we set up the fuse, detonator and suchlike on a very small scale. We lit the fuse, threw it in the pool and stepped back about ten yards. There was an almighty bang, and we were left standing on rocks, with no water in the pool and no fish to be seen anywhere. We spent the next two hours with someone chasing us in the forest. They never caught us!

The foreman on the Tighnabruaich Road gathered a few men around his Land Rover one day to give us a demonstration on electric detonating (for

12. Kyles of Bute from Tighnabruaich Road Viewpoint (with the big drop!)

blowing up rock for the new road). A burning fuse was the method used before.

We parked our machines well behind his Land Rover and went over for the demonstration. The foreman had a glass eye (he never told us how he lost his eye, it was probably from working with explosives). He had already wired up the explosives and had run the two wires back to his Land Rover. He explained how he would attach the wires to the battery and told us to go back behind the machines. When he attached the wires to the battery, there was a loud bang, followed by a shower of rocks. We peered out from behind the machines and saw the foreman diving under his Land Rover. When the dust had settled, we saw the foreman crawling out from below his Land Rover to survey the result, a bashed-up Land Rover with no windscreen.

Back in those days, most Friday nights there would be a dance, mostly Scottish music somewhere in the area, somewhere in the area meaning within approximately thirty miles distance. The boys would stand on one side of the dance, hall and the girls would be on the other side of the hall. Some of the girls would dance among themselves and the boys would eye up the girls, ready for the last dance. Early on, before the dance would start, most of the boys would purchase the courage bottle (a half bottle of whisky). Some of us back then were underage (for buying alcohol), so we used to get the older boys to go into the hotel to buy the courage half bottle. As we were not allowed alcohol in the dance hall, we used to hide the half bottle outside in a hedge or behind a wall, having a drink out of the courage half bottle every now and then as the evening went on. Then at last the time would come when the band would announce, 'Take your partners for the last dance,' and off we would go across the empty dance floor, having drunk the courage half bottle, heading for the girl you had been eyeing up all night. A lot of the time the girl who you had fancied all night long would turn you down and then you had the shameful walk back across the empty floor to where the boys were all cheering at your misfortune.

Down below where I lived was a small village called Newton. In the old days, a small ferry operated from here, across Loch Fyne to Furnace. In my day it was where we spent most of the time, out in boats and playing shinty in a clearing up in the woods.

It was getting towards the end of the summer, and time for the boats to come out of the water. One of our friends called Jimmy was about to take his boat out of the water when we all decided to go for one last cruise across Loch Fyne to a hotel called Minard Castle (we were just the legal age for drinking alcohol). I think there were about five of us on board, as the boat was a fair size, and had a small boat getting towed behind, in case we had any problems getting ashore. Off we went, arriving sometime in the afternoon. As the tide was in, we anchored quite close to the shore and went ashore in the small boat. After a few hours of partying we thought we better head back to the boat. To our horror, the tide had gone out beyond where the main boat was, so we were stuck. As there was nothing else to do until the tide came in to float the bigger boat, we decided to go back to the hotel and party a bit more.

Eventually, the tide came in and we managed to get the boats back in the water. By this time, it was starting to get dark and the engine on the big boat would not start. After about an hour working on the engine without success, we decided there was nothing else for it but to tow the big boat with the small boat, as the loch is about one and half miles across at that point. Being in good spirits and having only one life jacket between us (not good out in the water), we came up with a plan that we would take turns in the small rowing boat towing the bigger boat, with the only life jacket to be worn by whomever was in the small boat. It was a long and dangerous crossing, as the wind had picked up, the waves were quite big and we were blown off course down the loch. After about four or five hours, we made it back across Loch Fyne, but were a long way from any houses to get a lift up the road. After a long walk, we arrived at Inver Restaurant, which still had lights on, so we all went in to receive a real telling-off from the owner, as the alarm had been raised that we were missing and the local volunteer coastguard were on alert in case we never turned up. The volunteer coastguard were phoned, and everything returned to normal. A few days later, Jimmy eventually got the boat started and returned up the loch to Newton.

One other thing I remember from that time was that we were made to go down to Newton on a Sunday morning to a place called Sunshine Corner. This consisted of a bell hanging from a tree, and one or two benches to sit on. It was an open-air Sunday School organised by a lady called Mrs Beaton. The only thing I can recall from that time was Mrs Beaton ringing the bell!

13. Newton Village from where we set off in the boat

London

When I left the road construction, I went to London for about a year to work in construction. I worked with an Irish construction firm who were working on Ronan Point, where a big gas explosion blew the side out of a block of flats. One man, who was in bed when the explosion happened, was blown out of his flat from the thirtieth floor. He and the bed survived the fall!

The IRA was active in London at the time. I worked with mostly Irish men. I got on great with them. Being the youngest on the work site, my job at morning tea break was to go to the local butcher and bakery and get steaks and bread for the Irish workers and myself. I would cook the steaks in a frying pan, cut the bread into thick slices, and then put the steaks in between the bread. That was all the food they had all day. We had the usual banter about the Scots and the Irish. I stayed in lodging with the Irish until one morning, outside of the house, there was a body lying on the pavement.

I left London because I got a bit homesick and needed to get back to the mountains. When I was in London, I think I only came home once by plane. I had my car with me in London, so when I left, I drove home up the motorway to Scotland. I can still remember the wonderful feeling hearing Scottish music on the radio as I crossed the border into Scotland. I must point out that I am half-Scottish and half-English, as my mother was from Norwich and met my father during the war. I started working again for the Forestry Commission, on deep draining and hill ploughing new ground, all over Argyll.

14. Forestry ploughing
Photograph courtesy Forestry Memories
www.forestry-memories.org.uk

15. Deep draining plough attached to two tracked tractors
Photograph courtesy Forestry Memories
www.forestry-memories.org.uk

16. The Paps of Jura

Kilberry

HILL PLOUGHING

Kilberry was definitely up there as one of the most beautiful places to work. It is known for the Kilberry Sculptured Stones. These stones are said to be from the medieval parish church near the Campbell of Kilberry mausoleum. The view from the mountain where we were working was outstanding! I could look across the water to the islands of Islay and Jura.

This area up on top was flat and boggy, with small peaty islands (I ploughed up lots of peaty ground). One morning it was misty when I drove up the forestry road to work. My workmate was there, sitting in his car on the road and looking a bit shaken. He said that he had been watching a man running around the peaty islands, ducking down, and then the man ran past his car and disappeared into the mist. My workmate was so badly shaken, because he thought he had seen a ghost, that he had to go home. We later discovered that the man had escaped from a psychiatric hospital in Lochgilphead and had come up over the hill to the area where we were working.

Another day, I was ploughing a long, sloping hill overlooking Jura. The sun was shining in the front window and I fell asleep while driving the tractor. The tractor went over an old turf wall and I was catapulted out through the windscreen, taking the windscreen with me. Fortunately, I got stuck in the headlights of the tractor; otherwise, I might not be here today—I might have been fertiliser for the trees! That was just the start of my problems, as the tractor was still travelling down the mountain with me lying on the bonnet. I had to wriggle back along the bonnet, get inside the travelling machine and stop it, which I managed to do. Maybe I had been dreaming of having a beautiful dram of Jura whisky, or had had one too many drams the night before.

17. Forest at Lochgoilhead

Lochgoilhead

DEEP DRAINING AND PLOUGHING

When I worked around Lochgoilhead, I was ploughing and deep draining hills. I was working in Lochgoilhead at the time of the Cold War. Often, from the top of the mountain, we could see the nuclear submarines coming and going on Loch Long. One morning, the weather was very wet. My workmate and I were sitting in the car waiting for the rain to stop, as we had to walk a bit of distance to get to the tractors. I suddenly remembered that the night before I had been given bottles of whisky which were still in the car with us. The man had started a whisky company near where I lived in Glen Coe. He had asked me to deliver the bottles to one of the hotels near where we were working. This man also gave me a half bottle of whisky for myself. As it was so wet, we were tempted to sample the whisky, as the rain kept falling, but willpower held out and we sampled the whisky after work. In those days there was no breathalyser test before you started work.

We were also working on the opposite side of the loch. I remember it to be quite steep. At the bottom of the hill, there was a big flat area between the road and the loch. Another tractor driver working there took his tractor down to the loch to wash it when the tide was out. He had to travel a bit to reach the water. The tractor became bogged down and he was not able to move it. It was a Friday, so he left it there. Over the weekend, the water went over the tractor a few times. On Monday morning, the tractor driver phoned the Forestry Commission headquarters at Chapelhall near Glasgow to let them know that the tractor was bogged down. Some of the bosses appeared at the site. They asked him where the tractor was. The tractor driver picked up a stone as the tide was in, threw it out into the water and said the tractor was out there somewhere. Rumour was that he got fired on the spot.

18. Ben Cruachan from Fearnoch

Fearnoch and Taynuilt

DEEP DRAINING

I cannot remember too much about Fearnoch and Taynuilt, except that there was a very large flat area that had been ploughed sometime before we got there, and the furrows were full of water, like a big loch. I was working with another man, Archie. We set off across the flooded furrows. I was in the lead tractor, with the water following behind Archie who was on the other tractor with the deep draining plough attached. We headed for a narrow area where the ground seemed to slope away. As I was in the lead tractor, I went over the crest first and my mate Archie was coming behind, but we underestimated how much force would be in the water. I was fine, but poor Archie got the full force of water and the soil from the furrows in his back window and nearly drowned.

Peter Weir

I also ploughed up some charcoal pits which were quite common in that area because of the history of Taynuilt and the iron furnace located just north of there. At the time, I did not think about the charcoal pits, but now I realise the history of the charcoal pits and the iron furnace. The iron furnace was working from the mid-1700s to the late 1800s. The iron furnaces used charcoal from the pits.

19. Glen Creran
where my friend's grandfather's house was (now a ruin)

Glen Creran

DEEP DRAINING

In Glen Creran, I was deep draining high up on the mountain and lower down in the valley. Deep draining was to stop the trees from blowing over, because the Forestry Commission thought that by deep draining, the tree roots would go down deeper. It was never really that successful, as one can see if you go and look at a windblown forest and see the big flat root plate, the reason being the wet climate in the west of Scotland.

Glen Creran was another one of the beautiful places where I worked. High up on the mountain was a place called Bullough's Loch, which was set in a hollow surrounded by mountains, and was probably one of my favourite areas.

One Monday morning, we were driving up the forestry road in the Land Rover which we used to get to work, as the road was steep and a bit rough. There at the end of the road was a car and caravan, and the caravan was in the ditch. Two elderly people emerged from the car, and were so glad to see someone as they had not a clue where they were. Back then there was no satnav or good phone reception in remote areas. They thought they were on the main road to Glen Coe. As it turned out they had been stuck there since Friday night, when they tried to turn the caravan and ended up in the ditch. We proceeded to tow them out of the ditch, and checked to see if the vehicles were okay. Everything seemed fine, so off they went, never to be seen again.

The Forestry Commission planted trees all around that area. We deep drained all around Bullough's Loch, crossing over the head wall on several occasions. We continued to deep drain the surrounding area which was a stunning glen. A few years later I heard that the Forestry Commission had blasted a hole in the dam wall and drained the loch because they thought it was unsafe. Later I heard that the loch was drained because men were going over the hill from Ballachulish to poach deer and fish. The Forestry Commission said that their activities in the area were a fire risk.

On one side of Bullough's Loch was a beautiful corrie. One day while working, we thought we heard gunshots. Later that day, the gamekeeper appeared with traces of blood over his arms. He said he had shot one or two deer and asked us to take them down the road. We had a trailer which was used for taking fuel out the hill for the tractors. We hitched up the trailer and proceeded up to the corrie. To our horror, we discovered that the gamekeeper had shot eleven stags. That corrie lost its beauty that day. Some gamekeepers had a reputation for cutting fences to let deer in for shooting. When I was building fencing for the Forestry Commission, I repaired the cut fences.

At the head of the glen, below Sgùrr na h-Ulaidh, is an old ruin of a house. A good friend of mine, Walter, said his grandfather was a shepherd and lived in that house. He and his family, with all their possessions, left the glen and travelled over the steep pass into Glen Etive and onwards over the Devil's Staircase in Glen Coe to live in the area of the Blackwater Reservoir. What a challenge! Their home was where the reservoir is now, so the house is under the water.

While were we working in that area, someone told us about a cave not far from the road up the side of a small stream and under an overhanging tree. Neil (my workmate) and I decided to check it out. We found the entrance of the cave, which was quite tight to get into, and proceeded to crawl along the damp cave. It went for about fifty yards and then came to a dead end. We thought at that time we might have been the first to have gone so far in; however, the torchlight shone onto a forestry rubber glove lying on the floor of the cave!

The first *Highlander* film was made in Glen Coe, which is just over the hill from Glen Creran. I worked on the film set for the castle; this is the location for the swordfight between Christopher Lambert and Sean Connery. I also worked on another scene, where the Highlander lived in a cottage; he stayed young and his wife grew old.

I helped to build the castle where the sword fight was filmed in a thunder and lightning storm. I was also asked to cut the top of the castle down after the thunder and lightning storm with my power saw as it was made from wood and plaster. When the director asked if I would cut the top of the castle down with the power saw I replied, 'I will wreck my chain on the power saw with all the nails in the wood.' He replied, 'Peter, even if I have to buy you a new power saw, it will cheaper than paying Sean Connery to hang around.'

As the director knew I could work a power saw, he asked me if I would make up the gallows for the bodies (dummies) to hang from, which I did. That scene is near the start of the film, where the Highlanders are marching towards a castle.

20. Michael Palin, son Tom and Ang Phurba (Nepalese Sherpa), Aonach Eagach Ridge
Reproduced by kind permission of Hamish MacInnes

21. Michael Palin, his son Tom, me, and Ang Phurba
(I am the one with the moustache!)
Reproduced by kind permission of Hamish MacInnes

The Life of a Forestry Ploughman and Other Adventures

When I was in the Mountain Rescue Team, Hamish MacInnes was the leader; he made a number of climbing films at that time. When he was filming in Glen Coe, he used to get some of the team to help out. He was making a film with Michael Palin, Michael's son Tom, and a Sherpa from Nepal. The plan was to film them going along the Aonach Eagach ridge in Glen Coe. We spent a wonderful day up on the ridge with Michael, Tom and the Sherpa and with Michael's sense of humour it was great fun from start to finish. I forgot to mention—my job was carrying the tripod for the camera.

A few years later, when we had the River Coe Restaurant, Michael appeared, with most of the *Monty Python* crew. They were revisiting some of the areas where they filmed *Monty Python*. The Bridge of Death scene was filmed up in the valley of Glen Coe, and some of the mountain rescue team worked on that scene. The other place where they filmed was Castle Stalker in Appin (Castle Aaargh), situated in Loch Linnhe in a stunning area and below the Appin hill where I spent some time working with the Forestry Commission.

Michael asked my wife, who was the chef, if she would make two large apple pies for the Monty Python crew, as they would be coming back later after visiting their old film location. Michael and the crew arrived back and succeeded in eating the two large apple pies. Michael noticed the photo hanging on the restaurant wall, and we had a wee chat about the day on the ridge and then he signed the photograph for us.

22. Castle Stalker, Appin (Castle Aaargh)

23. Michael Palin, his son Tom and Ang Phurba on Aonach Eagach Ridge
Michael signed this photo for my wife and me.

24. Robin's Farm in Appin looks out towards Castle Stalker

Appin

DEEP DRAINING AND CUTTING FIRE BREAKS AROUND FOREST

Appin is another beautiful area. In Appin, I was deep draining and swiping (cutting the heather and long grass) around the forest for a fire break. In the mornings we passed the farmer's house, and every morning we would be called in for a cup of tea. Robin and his wife were real Highland people, with a big tea bill!

The forest at Appin was small, but had the biggest anthills that I have ever seen. The anthills were about three to four feet high. The rock in that area was limestone. That area had some deep holes like sinkholes and some holes like caves. When I was in the Glencoe Mountain Rescue Team, we went down a hole after some potholers from England discovered it. It was not my favourite place. Since then, one man has died in that pothole.

We travelled up Glen Stockdale in Appin with the Caterpillar tractors for a few mornings. One morning, on the track we travelled up, a great big hole developed in the ground because of the limestone and the vibrations of the tractor. The hole was big enough to swallow up the whole tractor.

Another day in Appin, the local forester asked myself and my workmate Neil if we would help Robin, the farmer, to burn a fire break up on his hill close to the forest, near a loch where people go fishing. We burned an area along the side of the forest to stop any fire spreading into the forest, and put out the fire, or so we thought. We continued to control burn Robin's heather on his hill and were about half a mile away from the forest fence when we heard him shout, 'Look behind you!' The area beside the forest that we control burned had caught fire again and was burning inside the forest fence. The three of us ran back as fast as we could to the fire. By this time the fire was raging inside the forest fence. We all had fire beaters with us, and after about half an hour, with sweat pouring off us, we finally managed to put the fire out. That is what we called a close one.

When we were having a cup of tea one morning, Robin noticed I was wearing a new Gore-Tex jacket. He said, 'I like your jacket. Where did you get it?' and I replied, 'Fort William.' After checking out the jacket he said, 'I will need to get one.' As I was going up to Fort William the next day, I said to Robin, 'Give me your size and I will get you one.' A few days later, when I was passing the farmhouse, I gave Robin his new Gore-Tex jacket. He was delighted with his new jacket and tried it on. It fitted perfectly and Robin said, 'Thanks Peter, I will get some money and pay you later.' After a few days, we had finished working in that area and moved to work in Glen Coe. With moving to work in another area, I never saw Robin for about fifteen years or more. One night I got a knock on my door in Glen Coe and there was a man standing at the door. The man said, 'Sorry, I am a bit late, but here is the money for the jacket, and I have added a bit of interest on since it has been some time since I got the jacket.' It was Robin. I hardly recognised him as it was years since I last saw him.

25. Hamish MacInnes filming a climb for the BBC
Reproduced by kind permission of Hamish MacInnes

Ballachulish and Glen Coe

DEEP DRAINING

Glen Coe and Ballachulish are two of the most beautiful and historic places in Scotland. There are so many stories from this area, but I will do my best to try and remember some of them.

When I was the manager at the Glencoe Ski Centre, there was the daily running of the ski centre in winter with all the normal problems: weather, roads blocked, staff, snow conditions which could change by the hour, skier safety, and so on. On one occasion, we had a massive snowfall and it was just before a main holiday. BBC Scotland phoned, wanting to come and film the snow, and to interview me. They duly appeared and did the interview with me and filmed the good snowfall. The film went out on the BBC news nationwide. A day or so later the weather changed (as it does in the mountains), and we had a wet and mild spell of weather. As that weekend was the holiday weekend, lots of people from all over Britain turned up to go skiing, not checking the up-to-date weather forecast for the ski centre. The ski hill was in a poor state after the heavy rain and the mild spell. There were a lot of unhappy people around that day. The snow conditions for that day were posted at the base station, as usual, so that people could decide to purchase a ticket or not.

During my time as manager, the Glenshee Ski Centre bought the Glencoe Ski Centre and they were prepared to put a lot of investment into the Glencoe Centre. Glenshee brought a bit of life back into the ski centre and they were great to work for. The first thing they did was to take us into the modern world, and purchase a brand-new large snow groomer which transformed the whole mountain and made life a lot easier for me. Back then, I did all the snow grooming, with an early start in the morning and, depending on the snow forecast, working late after all the skiers had gone.

We were planning to build four new tows, one new chairlift across the plateau, one big restaurant on the plateau to replace the old one, and one small one at the top of the access chairlift, mostly for summer trade. At the bottom of the mountain, we were going to add more car parking to the existing car park for the extra skiers, as the existing car park was not able to cope on a busy day. We were also planning to increase the size of the base station café and add on a new building for some accommodation for staff, as there was no security at night-time. The local planning and building authority, along with SNH (who had a big say in the development), turned most of our plans down. The directors of the Glenshee Ski Centre had a meeting and decided that, since the local authorities did not want any investment at the Glencoe Ski Centre, they would take their money somewhere else, which they did, and built an eighteen-hole golf course at Alyth over near Glenshee.

During a happier time, the BBC phoned up. As we had had a good fall of snow and the next day was Christmas, they wanted to come up to the base station and do a live broadcast for Breakfast News. We were one of the few places to have snow on Christmas Day. The crew arrived, filmed the snow around the base station and we did a live breakfast Merry Christmas from the ski centre.

When I was manager at the ski centre, I turned two of the old buildings at the top of the car park into a ski and climbing museum. It had a lot of good artefacts, as well as history on climbing and skiing in Scotland, including Sir Hugh Munro's ice axe. Sir Hugh Munro was the person who listed all the Scottish mountains over 3,000 feet. We had enough climbing and skiing material to fill the two old buildings. After I left, the museums were closed. I do not know what happened to all the artefacts.

One day, before I became manager, I was skiing at Glen Coe. The new police sergeant had just arrived and was skiing just below me. I did not know him too well at this time, apart from meeting him on mountain rescues, so I thought I would go and have a chat with him. He was about fifty metres below me, so I headed down the hill towards him, with the plan to stop and slide towards him. Unfortunately, I hit a large bump in the snow and went flying. I hit the new police sergeant about waist height, and one ski went through his police issue jacket and out the other side.

When I got myself out of the snow, I looked over at the police sergeant. His ski goggles were lying sideways on his head, one side was filled with snow and he had a funny look on his face. When we both recovered, all I heard was, 'Weir! I might have guessed.' We had a good laugh and have been good friends ever since. The police sergeant spent the rest of the skiing season sporting a big bit of tape over the hole where the ski had gone through the jacket.

While I was managing the Glencoe Ski Centre, I did the snow grooming and back in those days we had only a small groomer. One day, after a big snowfall followed by heavy rain and high wind, three of the staff and I went up the mountain to repair some of the tow cables that had come off the pylons in the high wind. We were heading up towards an area called The Canyon, an area which was never known to avalanche. Two of the men were on the back of the groomer and a man named Innes was in the front with me. Suddenly, the area in front of the groomer exploded and a massive wall of snow and water came towards us. I managed to lift the front blade up to hold back some of the snow and water. The snow groomer, with me and the men in it, was pushed about two hundred metres down the mountain at a terrifying speed, and buried by the avalanche. We ended up on one of the chairlift platforms, with myself and Innes completely buried. We could see nothing but snow. We sat and looked at each other for a few seconds and one of us said, 'at least we are sitting upright.'

After spending a few minutes trying to work out what to do, we saw some movement at the side windows. Luckily, the two men on the back of the groomer had only got buried up to their waist and had managed to free themselves. The two men managed to get a shovel from the hut near where we ended up, and they dug a tunnel into our window. As soon as we saw the shovel at the window, we managed to kick the window out, and to escape out the window.

The avalanche was so big and powerful it had snow boulders around the groomer as big as the machine itself. The avalanche continued down towards the next tow lift and demolished the lower tow hut, just leaving the metal pylon standing (it was concreted in). The avalanche continued across an area called The Plateau for about half a mile and then continued down the front of the mountain. That is what you call a near thing; if the snow groomer had flipped over, we would not be here today. I took the boys who were on the machine with me down to the Kingshouse Hotel for a wee dram to calm the nerves.

Network Rail had a communications station on Meall a' Bhuiridh at the top of the ski centre. It was run with large gas bottles, as there was no power on top of the mountain back then. Being so high up and with so much snow on top of the mountain, there were always some problems digging out the bottles, and digging out to get into the hut. The biggest and most expensive problem, as well as manpower, was transporting the gas bottles up the mountain by helicopter, as back then we did not have a big snow groomer.

Network Rail had a meeting with me at the ski centre to see if we could get power to the top of the mountain, as we had power midway up to the café area. I agreed, but had to check up with my boss who owned the mountain, as the power cable had to be buried to the top of the mountain, where possible. The boss agreed, and the deal was: if we dug the track and buried the power cable where possible, Network Rail would do the rest. It was a good deal for the ski centre, as all the ski tows in that area ran on diesel.

The best way to dig the track for the cable was to fly a small digger up to the summit and work down the mountain, as at that time there were problems with machines on the mountain. It was all go. We hired the helicopter and as the digger was too heavy to lift in one go, we had to take the digging boom off the machine, and fly the boom and main digger up separately. Everything went well, as we also flew all the power cables up the mountain, so they could be laid out by pulling them down the mountain.

I started digging the track from the summit down the hill, which had a lot of rocks all over the place. I got a call on the radio from base station to say that three of the Network Rail bosses were on the way up the mountain and they would meet up with me when they got there. The Network Rail bosses got so far up the mountain on two chairlifts, and they had to walk the last bit to the summit, as the last lift to the summit was a drag lift. I was digging away with the track, and getting on well, when I saw the three Network Rail men approaching about fifty yards away. I got a bit distracted and, as I was moving forward, the digger mounted a large boulder. Before I had time to stop, the digger flipped over, trapping me inside as it landed on my door. The three men shot over. As it was not a very big digger, and because of the way it landed, they managed to lift it back over onto its tracks. We had a bit of a laugh about the incident and a 'catch up' on how the job was going, and off they went. The job was completed, Network Rail got their power to the summit, and we got power to the rescue hut area and workshop.

Not long after I got the manager's job at the ski centre, we had a massive snowfall and there was so much snow we had to shut the top part of the mountain, as some of the pylons were completely buried under the snow, and there was no way of operating the lifts. There are photos somewhere of one of the ski patrollers standing on top of a pylon near the summit, and you cannot see the pylon under his feet. The ski patrol hut and all the middle tow stations were buried under the snow. We had a problem with holidays coming up; there were no top tows working on the top section of the mountain and the mid-section was buried at the top of the tow. We were in a bad situation. Nowadays that would never happen with the large snow groomers. There was only one thing for it, and that was to hire a large digger from a local contractor, which I did, and to travel the digger all the way from the car park up to the bottom of the top lifts. Sandy the digger driver did a good job, and dug out access to all the lifts and rescue hut and we got most of the hill operating, apart from one tow.

After that big snowfall we decided to replace the older pylons and put new higher pylons on that tow lift. We had these made by a welding firm in Fort William, then flown up by helicopter and dropped into place. It turned out

41

to be a big and expensive job, as we also had to put new bases in for the new pylons.

One day (I cannot quite remember how it came about), we were offered a trial use of a skidoo that had last been on hire on a James Bond film (or so the people offering the use of it said). It was a beautiful machine. It was black with gold writing on it, and great to drive. It was Easter, and the ski centre was full of skiers and snow boarders. Dave, the ski patroller, and I thought we would take it for a drive down from the rescue hut to the plateau café which is at the base of the cliff-hanger chairlift. As it was a beautiful sunny day, we had our sunglasses on we thought we looked cool sitting on the nearly new ex-James Bond skidoo. Off we went, with me driving and Dave on the back, down to the plateau café. As we got to the bottom of the cliff-hanger chairlift where we were going to park it, there was enough snow to get to the café and there was a large queue for the lift. I thought I would be cool and slide sideways to stop at the side of the queue, which I did, but was going a bit too fast and the skidoo flipped over and trapped Dave and myself under the skidoo. The whole queue of skiers and snowboarders cheered and burst out laughing. Some came to our help, and lifted the skidoo off us. What an embarrassment!

One Sunday, on one of our busiest days, we had great snow cover and good weather, apart from the wind that was picking up. It was the afternoon and I was up on the summit when I got a call from the operator on the bottom chairlift to say there was a noise coming from the gearbox on the access chair. I radioed the base station back, saying that I would make my way down to the top of the chairlift. On the way down, the chairlift operator radioed me back to say the noise was getting worse. As we had a backup engine but no backup gearbox, I sent out a radio message to all the lift operators and ski patrol to close all the lifts, and woman and children only to go on the chairlift on every fourth chair in case the gearbox seized up, so that there would not be too many people on the chair.

When I got to the top of the access chairlift, the women and children were queuing to go down. Luckily, most had made their way to the chairlift and were on their way down. On the other hand, the men were all complaining about having to walk down with ski boots on, as there was not enough snow to ski all the way down. The operator at the bottom was keeping me informed about the noise from the gearbox, and it was getting worse. I had a serious problem, so I told the two top lift operators to put the barrier down as soon as all the woman and children were on the chairlift, and not let anyone else on. I made my way running down the mountain. When I got to the bottom of the chairlift, the operator said it was getting bad and hot; fortunately, the women and children were nearly down. The last chair arrived at the base station with a woman and child on board and I was a very relieved man.

By this time, the gearbox on the chair lift was red hot and making awful load noises, and all the male skiers were starting to appear at the base station wanting their money back and really whingeing about having to walk down the mountain wearing their ski boots. I was not too worried about the whingeing men (it was in the afternoon and most people had had their money's worth); I was just glad to get all the women and children off the chairlift, as lowering people off the chairlift would have been a major rescue operation, especially with the wind picking up.

All the season-ticket skiers (regular skiers at Glencoe) walked down: men, women and children just took it in their stride. The ski patrol were excellent—people you could rely on in an emergency.

The outcome of that day was: we had to get a new gearbox, and the chairlift was out of action until the gearbox was replaced. If something similar happens today they won't have the same problem, as there is a road from the carpark up to the top of the chairlift, an excellent safety feature if the weather turns bad, which it can do very quick on the mountain. The unpaid weekend ski patrollers were amazing. They were all from the Glasgow area, fully qualified, trained in first aid, and some were also doctors. They helped whenever there was a problem: injuries, digging out tows and so on. Back then we could not have functioned without them. Thank you, old ski patrol.

Back then the local police were different. They were part of the community, and got involved in the goings-on in the village. One of my favourite memories was when the police sergeant met me in the village and said, 'Peter, as you work in the forestry, can you get me a Christmas tree?' As he was a good friend, I replied, 'do what I do: go up into the forest and get one.' That forest area had lots of small trees growing everywhere (it was disputed land).

About a day or so later, I happened to look out the window. I saw three men, one with a saw over his shoulder, heading up into the wood above the road: friends of the policeman! I popped on my wellies and headed for the wood above the road. I hid behind an old stone wall, and as the three men approached carrying Christmas trees I shouted, 'Hey! What do you think you are doing?' They dived for cover and popped their heads up like startled roe deer. The men were not in uniform, and nobody was hurt in this incident. Great fun! What has happened to the old Scottish way of life? Give me the old days any day!

BACK TO DEEP DRAINING

When we arrived in Glen Coe, we unloaded the tractors and plough, and we were met by the local forester at a place called Inverigan. In front of us we saw an old ruined house. The forester asked us to knock down the walls as it was unsafe. We knocked the walls down and I did not know the history of the area. To my horror, I learned later that it was one of the areas where the Massacre of Glencoe had taken place in 1692. The house at Inverigan was one of the main sites of the massacre, where nine people were murdered. A memorial for those murdered in Glen Coe still takes place every year on 13 February. A service is held at the church in Glen Coe, and then there is a walk to the Memorial Monument in Glencoe village.

26. Old Inverigan House in Glen Coe before the walls were knocked down
Nine people were murdered here in the Massacre of Glencoe

We started to drain the area behind the house ruin and saw a few more ruins. This area became the forestry campsite and now it is a camping and caravan club site. The area where the manager of the forestry campsite stayed is now the new National Trust for Scotland visitor centre. There is still the ruin of a very old sheep fank in the area behind the Inverigan house ruin.

In Ballachulish and Glen Coe, I did deep draining. It was winter when we arrived in Glen Coe. It had already been planted and the trees were about four or five feet high. We deep drained the front of the mountain and then went over the top into Ballachulish, which was famous for its slate quarry. Coming into Ballachulish there were some small test quarries on the way down and there was also an old farm ruin called Brecklet which was hidden by the forest around it. My friend, Walter, who was a shepherd, remembers having a meal in that house. At the bottom of the forest there was another old quarry which had a big, deep hole filled with water. The locals used to dispose of their old cars there. As I had a dosing bar on the front of my machine for pushing trees down, the forester ordered me to get rid of all the vehicles into the hole.

27. The old farm ruins at Brecklet
My friend Walter remembers having a meal in the house.

28. One of the ruins at Brecklet

On my way back over the top into Glen Coe, we were hit by a big snowstorm and could not see through the windscreen. We had to leave the machines on the mountain and walk down. It was about three weeks before the snow melted, and we were then able to get back to the machines. Most of the Glencoe forest has been felled now and some of the Ballachulish forest that remains is also getting felled. I fell in love with Glen Coe and decided to stay.

29. The Glencoe Mountain Rescue Team in the 1980s.
(I am in the front, with the moustache, and wearing black cap and red jacket)
Reproduced by kind permission of Paul Moores

I was in the Glencoe Mountain Rescue Team for about twenty-one years and enjoyed being part of the team. The Glencoe Rescue Team began with Walter Elliot (Senior), who lived at Achnambeith with his wife and family. His son, Walter, went on his first rescue with his father when he was 16 years old, along with local shepherds. In the nineteenth century, another local shepherd, Nichol Marquis, who lived in the farm below Ossian's Cave, was the first on record to climb into it. My daughter is a descendant of his. Later, the team was formalised by Hamish MacInnes and local people who had an interest in the mountains.

30. Walter Elliot and sons, Walter and Willie, about 1946.
Walter (centre) still does the call out for a mountain rescue.
Reproduced by kind permission of Walter Elliot

During my time living in Glencoe I have also worked with some television and film companies. These projects include climbing films, the feature films *Rob Roy*, *Highlander*, and television programmes, including *Blue Peter*, BBC News, BBC documentaries and BBC films.

I met John Smith when he was the leader of the Labour Party and Shadow Chancellor. We spent a wonderful day filming in the mountains in Glen Coe with him and some of his friends from parliament. John Boyce was the one doing the filming. I got to know John and he asked me to carry some of the film equipment and do the sound on the mountain.

John Smith and I discovered when we were on the mountain that we both went to Dunoon Grammar School. A week or so later, John sent me an invitation from the Houses of Parliament for a Dunoon Grammar School reunion. I did not attend the reunion, as I never did as well as John did at school because I left at about the age of fifteen.

John Smith was a very witty man and there were a lot of jokes being told. As we were approaching Stob Coire nan Lochan we looked across the valley and someone said, 'John, that mountain over there is called The Chancellor. John replied that they should take the present chancellor up there and push him off (the Tories were in power then)! A bit further up the mountain, we turned around and saw two men coming up the path; they were stripped to the waist, as it was a warm day. They looked like body builders. John looked around and quick as a flash said, 'I hope they are not Tories!'

The whole party were extremely fit, so we proceeded up the ridge onto Bidean nam Bian, the highest mountain in Glen Coe. We sat on the top and had some tea and snacks and had more good crack. After we had rested, we proceeded down the ridge into The Lost Valley (Coire Gabhail) and back to the road. It was a great pleasure to have spent a day on the hill with such a fit and wonderful man.

The Life of a Forestry Ploughman and Other Adventures

31. John Smith and myself, the author

32. John Smith and his friends from parliament, with cameraman John Boyce

Peter Weir

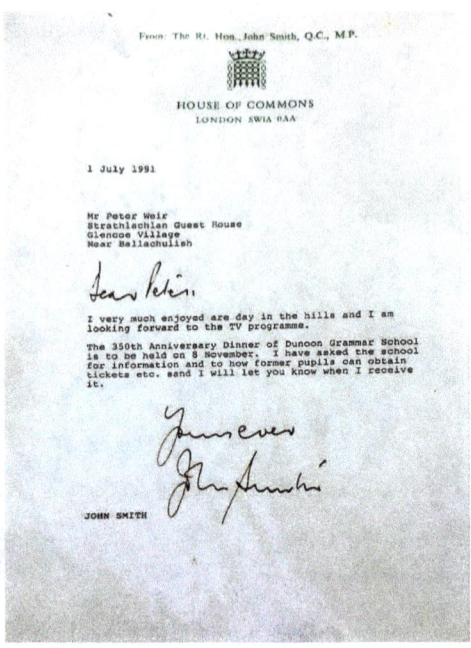

33. Letter from John Smith about the Dunoon Grammar School reunion

34. Glencoe Ski Centre

Left to right: Philip Rankin, who started Glencoe Ski Centre and unfortunately is no longer with us; Doc Maclaren, Ski Patroller; I am the one with the moustache; Paul Moores, famous climber and member of the Avalanche Information Service
Reproduced by kind permission of Paul Moores

35. Paul Moores and myself at the Glencoe Ski Centre
Reproduced by kind permission of Paul Moores

36. Rannoch Moor

Rannoch Moor

DEEP DRAINING

I remember this area being a small forest. It belonged to an estate and was planted as a shelter break for deer. Because there were very few native trees in this area, the deer could shelter in the planted forest during bad weather.

That area was full of old roots forming part of the old Caledonian forest which covered most of Rannoch Moor many years ago. The old Caledonian roots were as fresh when we burst them open as they were when they were growing.

I was told by the ecologist Dr Adam Watson that the climate had changed years ago, and a very wet spell in this area created poor conditions for the Caledonian pine.

We deep drained this area to stop fires spreading, as the hill lochs on Rannoch Moor were a popular place for fisherman coming from the towns.

Those fishermen would light fires to boil their teakettles, and sometimes a fire would spread. The forest was not planted again after it was felled.

The forest at Rannoch Moor was up near the ski centre and this reminds me of the time when *Blue Peter* were wanting to do a film with some vision-impaired young people, and search and rescue dogs, up at the ski centre. I was not the manager of the ski centre at that time, but was a member of the Glencoe Mountain Rescue Team, so the *Blue Peter* crew asked if some of the team would be involved.

The plan was to take the young people up the chairlift and walk for a bit to find deep snow, so that we could bury someone in the snow and the dog would find them. As it turned out, I was the one selected to be buried in the snow. Some of the young people wanted buried in the snow so that the rescue dog could find them as well.

We dug a big hole in the snow. I climbed in, was buried, and waited for the dog to find me. As I could not see, I was told—later—that the dog searched around and sniffed out where I was, lifted his leg, and urinated on the snow above where I was buried. Seemingly, everybody had a great laugh at this and started to dig me out. As I appeared out of the snow hole, the dog jumped in and bit me on the arm.

The payment for this was that the Glencoe Mountain Rescue were on the front of the *Blue Peter Annual* and I was on the first page inside, with my friend the dog. I also got the famous *Blue Peter* badge.

One day, my friend—the police sergeant with the hole in his jacket after our meeting on the ski slopes—and I were going fishing out on the hill lochs on Rannoch Moor. We met Ronnie, who was in the Mountain Rescue Team with me, and asked him if he wanted to join us on our fishing trip and he replied, 'that would be great'. We warned him to take plenty of food with him, as we usually stayed out fishing for most of the day. Ronnie being Ronnie said, 'When I go fishing, I never take food with me.'

We headed off late morning onto Rannoch Moor and out to the loch where we were going to fish. The police sergeant and I always had a competition to see who would catch the first fish, so as soon as we reached the loch I put my fishing rod together and cast out onto the loch. Straight away I hooked a fish, and when I reeled it in I noticed that I had forgotten to take the plastic cover off the hook. After a few hours fishing, I met up with the sergeant for some tea and sandwiches and proceeded to sit down at the side of the loch to enjoy our picnic. We were both watching Ronnie, who was down the side of the loch fishing. When he saw us sitting down, he started to work his way up towards where we were. Ronnie looked at us eating our sandwiches and drinking our tea. After a few minutes he said, 'I am starving. Any chance of some food?' We always took plenty food and tea with us, so Ronnie got fed.

Another day, Ronnie and I were fishing in Loch Achtriochtan in Glen Coe when a storm blew up, and we were blown back up the loch onto the farmer's field at the far end. We had a few attempts at trying to row the boat back up the loch, failing every time, and ending up on the farmer's field. We gave up, tied the boat to a fence post and walked back to tell Willie, who looked after the boat, what had happened, and he had a good laugh at us. We never caught any fish that day, and had to wait until the wind dropped the following day to recover the boat.

37. On the road to Bonawe
This is what some of the hills would have looked like, with natural trees and animals grazing (without the mast on top of the hill!)

Barcaldine and Bonawe

DEEP DRAINING

Although I cannot remember too much about working in the forest at Barcaldine, I do remember the stories my bed and breakfast host, Clarkie, would tell me in the evening. One story still sticks in my mind. It was during the war when coal was short, and the railway ran in front of his house. One day, as the train was passing, he noticed there were some rabbits in the field, and the guard on the train was throwing lumps of coal at them. He clicked on a good idea: he got some stuffed rabbits and put them in the field in range of the train and, as it passed twice a day, he would go out and collect almost a bucket of coal. The stuffed rabbits were moved into a different position every day. He said the stuffed rabbits kept him warm all winter.

The railway track went from Ballachulish to Oban. Before the railway was closed in 1966, it was one of the most scenic journeys in Scotland; it was possible to start in Oban with the bus to Taynuilt, take the boat up Glen Etive, followed by the bus up stunning Glen Etive to Glen Coe, then down the beautiful valley to Ballachulish, and finally take the train back to Oban along the side of Loch Linnhe.

Nowadays, all the railway line has gone. The old track has been turned into a cycle route that starts in Ballachulish and goes to Oban. Progress? I think not! The traffic on the A82 has been increasing every year. In the height of the summer up in Glen Coe, you can get as many as fifteen to twenty buses at a time in certain beauty spots—not counting all the other vehicles.

Bonawe was an area up the side of Glen Etive, and we had to go past Bonawe Quarry to reach where were working every day. Back then, the road past the quarry was gravel. It went up the side of Glen Etive for some distance but unfortunately it did not go all the way to the top of the loch. There is a rough walking track along the side of the loch to meet up with the Glen Etive road. We were deep draining that area and the forest had already been planted. Nowadays the forest has expanded since we were there, and you get a good view of it from across Loch Etive at Taynuilt.

38. Approaching Tyndrum from the A85

Tyndrum

DEEP DRAINING

We worked in this area on both sides of the A82 road, with the railway on both sides. The local forestry men used to have a nickname for one of the local foresters—Profumo—as he wore so much aftershave; if the wind was blowing in the right direction, you could smell him coming long before he appeared. It was also a beautiful area. When I worked there, it was a small village, but today it is one of the busiest areas on the A82 road. The forest has expanded since my day and now there is also a gold mine and the West Highland Way, a walk that starts in the outskirts of Glasgow and finishes at Fort William.

As Tyndrum is on the main A82 road leading to the ski centre, it reminds of another story. It was summertime, and we always opened the bottom chairlift for tourists. One day, a few people turned up at the car park and asked to speak to the manager. I replied, 'That is me. Can I help?' They said they were making a film in the area and would like to use the Glencoe Ski Centre as one of the film locations. One of the gentlemen introduced himself as Michael Caton-Jones, the director. They wanted to shoot part of a film up on the ski hill. The film was *Rob Roy*, which turned out to be a blockbuster film. They wanted to go up the mountain to check out an area where they could shoot a scene with standing stones. I took them up in the chairlift to an area that I thought was suitable for what they wanted, and they were delighted with the location because of its outstanding scenery. As it turned out, the part of the film with the standing stones was shot in the mist, so the scenery did not matter.

After returning to the base of the chairlift, they agreed to use the ski centre to shoot the standing stone scene and there would be a payment for all the work and trouble. I told them I would have to check with my boss, as he owned the estate the ski centre was on. I phoned up my boss and told him what might happen, with his permission. He agreed to let them film on the mountain, as long as they were finished by a certain date, as he and his friends go shooting on the mountain (at that time of the year the ski centre was closed to the public). The director said that would be fine and that they would be finished by that time. The film crew asked if I could I get the standing stones transported up the mountain to the area we agreed on, if they delivered them to the car park. When the standing stones arrived, there were quite a lot of them, and some were too large to go up the chairlift. I had to hire a track machine to transport the large stones up the hill. The stones were made from plaster, so they were not too heavy. There was a bit of work involved getting the standing stones into position and securing them with ropes to stop them blowing away.

Weeks passed and there was no sign of the film crew. Finally, with one week to go before the deadline, the film crew turned up in the car park. They filled the car park with vehicles, horses, solders, Highlanders in kilts and so on. The director and some of the crew wanted to go up the mountain, and the assistant producer asked if he could use my office for himself.

It was all go, or so I thought, but I was in for a bit of a shock. The director and his crew came back down the chairlift and met me at the base station. The director said, 'We can't film here today. We will come back in two or three weeks and film.' They had a problem, but never told me what it was. I replied, 'If you can't film here this week, you won't be able to use the mountain, as my boss goes shooting up there in the next few weeks.' The assistant producer said, 'That will not happen. Money talks.' I replied, 'It will not talk here, as my boss does not need the money,' and sent them on their way. When I told my boss the story, he said, 'Well done. Peter. They had plenty of time to film.'

When I was up at the ski centre, I did a short half-hour interview with Monty Don from *Gardeners' World*. We were talking about the all the films made in the area. The interview was on the mountain at the top of the chairlift and he was a wonderful man, just the way he is on his gardening programme.

Another short interview I did up at the ski centre on the mountain was with Alice Beer, a BBC television presenter who was nice and friendly. We went up in the chairlift and walked up on the mountain to some beautiful viewpoints, talked about the history of the area, and had some good laughs.

39. Looking towards Ben More, Crianlarich

Crianlarich

DEEP DRAINING

The forest at Crianlarich, which had a railway station, was at the side of the both the A82 and A85 roads. It was in another stunning location. This was another area where we encountered the old Caledonian pine roots. When we hit a root with the plough, the roots were so fresh and strong that the shear pin would snap and gave us about one metre to stop. The shear pin was a safety device to stop us wrecking the plough. We went through a lot of shear pins. Every time we broke a shear pin we would have to reverse back and lift the plough and start on the other side of the root. As we could not complete the new drain, it left a bit of a mess, and the forestry workers had to sort it out by digging by hand. Back in my day the forest was not very big, but it has expanded a lot since then.

40. Hamish MacInnes
Reproduced by kind permission of Hamish MacInnes

HAMISH MACINNES

Hamish MacInnes, OBE FRSGS, is a Scottish mountaineer, mountain search and rescuer, author and advisor. He is a friend of mine, and was the leader of the Glencoe Mountain Rescue Team. We were both building houses about the same time and we would help each other out where possible. Hamish would phone me and say, 'I need a lift with something heavy. Can you help?' and I would do the same. Hamish also had a tractor, which came in very handy for moving building material around, especially as my first house was up a narrow road where the delivery large lorries had problems. As we were both in the rescue team at that time, another problem was that whenever we mixed concrete or mortar, we would get a call out for a rescue (not all the time), and so we had to throw away a lot of mortar or concrete as some of the rescues could go on for some time.

Hamish was in Glen Coe, making a climbing film which was going out on live television. There were four climbers involved on two separate routes—very difficult climbs. A few of the boys who were on the Glencoe Mountain Rescue Team were working on the film, setting up filming platforms, laying out all the cables from the roadside up to the climb (with the helicopter—back then there was no modern technology), and looking after the safety of the filming crew. With so much filming gear on the mountain, the BBC wanted a nightwatchman on the mountain for the duration of the film. I was selected to be the first night watchman, and at the end of the day everybody left the mountain, some by helicopter and some on foot. The helicopter landing area was a small area on the edge of a sheer drop. At the edge of a cliff we had a small tent which was going to be my accommodation for the night. I was just getting familiar with my accommodation when I heard the helicopter coming back up the mountain. It landed on the small area, and the pilot was nodding for me to come over to open the door, which I did, and there, lying on the seat, was a large bottle of whisky. The helicopter flew away and left me with a nice wee dram for a long evening.

Out of the blue, it started to get very overcast. A thunderstorm was about to start and as the area had a lot of metal around (all the metal dishes and aerials), I headed into the tent and decided to open the whisky bottle. It turned out to be a big storm and quite a scary experience, so I had a few drams out of the bottle. After a few more drams I needed to go outside to the toilet and as I was contemplating my future, I looked over the edge with the big drop and thought I better be a bit careful here. As I turned to go back into the tent, I noticed a climbing rope laying on the tent floor, so I proceeded to measure out enough rope to get me safely out of the tent for the toilet and then tied one end to the tent pole and the other end onto my ankle, so that I would not walk over the cliff. I woke up in the morning with a slight headache and a rope tied to my ankle. The BBC live broadcast was a big success, with all the climbers completing the climbs.

In another climbing film made by Hamish MacInnes for BBC television, I played one of the bodies. It was called *Duel with An Teallach*. The film is based on the true story of three mountaineers who were climbing An Teallach, a mountain in Wester Ross. Two of the climbers said after they had done the climb that they would wait while the third man went to the summit. The two climbers started to get cold, and began to make their way down the mountain, but they fell. The third person arrived back to find his friends gone, and as he descended the gully, he spotted them roped together, unconscious, supported by the rope that was hooked over a large boulder. Most of the film is about the climber, Ian Ogilvie, trying to rescue them. We filmed the part where the survivor tries to lower his two mates down the mountain in Glen Coe, up in Stob Coire nan Lochan, in a gully called Left Fork Gully. The filming was very successful. One day it was too sunny to film. My friend and fellow rescue team member, Will, was also working on the film, and we both had skis with us. When the director said the helicopter was not doing anything on the film for a while and asked if we wanted use it and go skiing, we did, and had a fantastic time.

41. A film Hamish MacInnes was making about the Glen Coe area.
I am the one in the stretcher.
Reproduced by kind permission of Hamish MacInnes

Dunoon and Arran

In my teenage years (hippie times), some of my friends and I would go to Dunoon and get the boat to the island of Arran with a few pounds in our pocket and a rucksack with a sleeping bag in it. The first thing we did when on the boat was head to the bar for a beer or two to get us in the mood for our big weekend. We never had a place to stay, apart from the famous bus shelter not far from the pier in Brodick (this was where a lot of young people gathered); this gave us a roof over our heads for the night. There was only one problem about staying in the bus shelter: everybody had to leave by about seven o'clock in the morning, as the fire brigade would arrive and tell everybody to get out as they were going to hose out the bus shelter (as they did). The police also gave us a hard time with so many young people on the island. They brought reinforcements from the mainland to try and keep order on the island.

One thing I remember very well was, one night after a few beers, the police chased us—for some reason—and we scattered in all directions on foot. I took off running up a road, and as it was very dark and had no torch light, I jumped over a stone wall, thinking I would land in a field. To my horror it was a bridge parapet and as I was flying through the air, I thought I was never going to land. Luckily, I ended up on the bank of the river in amongst bramble bushes and small trees, with one leg of my trousers torn off and covered in scratches from the bramble bushes. The police did not catch me, but I had to spend my weekend with one of my legs of my jeans missing and covered in scratches from the bramble bushes. After a wild weekend, it was time to catch the ferry back to Dunoon, so we would head back to the pier in Brodick, with a few policemen in the area watching us to make sure we got on the ferry.

As we went to Arran on several occasions, we had a routine when the ferry was leaving the pier. With a few more policemen watching on the pier to make sure there were no problems, we used to go to the toilets and get toilet rolls, go up on deck, hold one end of the toilet roll and throw the roll like a streamer onto the pier (at that time we thought it was fun).

There is one other story which I remember from my Dunoon School days. Another friend (who also did not like school) and I used to go down to the shore on the Firth of Clyde and hire a rowing boat which we paid for out of our lunch money. We would go along to Dunoon Pier and take the rowing boat under the pier, and as the ferries were coming and going, they used to create very large waves, which was great fun at that time. At that time going out in the boat was better than standing in the corner of the room waiting to get the belt.

42. Tighnabruaich and Kilfinan area

Tighnabruaich and Kilfinan

HILL PLOUGHING AND DEEP DRAINING

It was in my early days when I worked in this area; it is a massive area now. It was good hill ground for sheep and cattle. I stayed in two places: one was the Royal Hotel, Tighnabruaich and the other a farmhouse in Kilfinan, close to Tighnabruaich. My father knew the owner of the Royal Hotel, so the deal I got was that I could stay free if I cooked my own breakfast. The other part of the deal was that as I was a resident in the hotel the local customers (the regular local customers) were my guests, so they could carry on drinking after hours, which was ten o'clock back then. Quite often we would end up in my room having one for the ditch, and quite often I would wake up and find people sleeping all over the room.

When I was staying in Kilfinan, I stayed with the local farmer in the farmhouse. As it was quite remote with not much to do in the evening, we used to go to Kilfinan Hotel, not far from where I was staying. The only place we could go for entertainment was the pub. The owner back then was a bit of a character. As it was a quiet time of the year, we would walk in and ring the bar bell, the owner would appear and we would get the usual greeting: 'It's yourselves. Just help yourself and put the money in the till,' which we did.

When I was ploughing there, the local forestry men were coming behind, planting in the furrows I had just ploughed. The forestry men were on piecework—paid for each bundle they planted. As the price they were getting was very poor, it was common to get rid of some bundles. Instead of planting on top of the furrow, the men were coming behind me and getting rid of lots of plants underneath the furrow.

Another story that comes to mind is about adders. Adders are the only poisonous snake in Britain and the Tighnabruaich area was one place where we ploughed them up. One of the forestry men had killed an adder and he thought it would be a bit of fun to put the dead adder in his workmate's piece box (sandwich box). He thought that his workmate made his own sandwiches, but actually it was his mother who made them. It was said that his mother opened the piece box and passed out on the kitchen floor.

On another occasion when we were deep draining, I was in the lead tractor and went over an old stone wall. I turned around and spotted a nest of adders. They started to move very slowly, as I think they were hibernating. I stopped to look at Archie on the other tractor and he was waving me back, so I put the tractor into gear and pulled Archie and the tractor and plough over the adders with Archie up on the seat of the tractor.

One other thing I can remember from that area was that we had to walk for some distance to get back to the road, as there were no forestry roads back then. I was heading down the hill on my own, with my piece bag over my shoulder after work, and was in the process of climbing over a farm gate, when suddenly there was an almighty bang which made fall over the gate. When I picked myself up, I thought someone was trying to shoot me, and I disappeared down the hill back to the road at a very fast speed. When I was having a meal in the farmhouse where I was staying, I was telling my story about the bang and the farmer said he had heard somewhere that the bang (boom) was a test flight for Concord, the new supersonic airliner.

Not far from the farmhouse, on the shores of Loch Fyne, is Portavadie, where the government was going to build oil rigs for the North Sea. A great big hole was blasted out of the ground and then flooded for the oil rigs. A whole village—Polphail—was built on the mainland to accommodate the workforce but, as it turned out, there never was an oil rig built in Portavadie. Polphail became a ghost town that was never used. Nowadays there is a beautiful marina where the hole was, and a beautiful hotel, the Portavadie Marina, with luxury accommodation and spa. A car ferry runs daily from Tarbert (not to be confused with Tarbet, Loch Lomond) across Loch Fyne to Portavadie, and it is a stunning location, definitely worth a visit.

One night when I went into the farmhouse, the farmer was at his sink with his hands and lower arms covered in blood, and had the contents of a sheep's stomach all over him. He said, 'Peter, I have a treat for you tonight. I have got tripe.' Tripe is part of an animal's stomach. I had one look at him and said, 'Sorry Neil, I cannot stand eating tripe.' I had never eaten tripe in my life before but, looking at the state of farmer's hands and arms, it put me off eating tripe for life. I think the farmer was a bit taken aback about my reaction to one of his favourite meals.

43. Tarbet, Loch Lomond

Tarbet, Loch Lomond

DEEP DRAINING

Overlooking Loch Lomond, Tarbet was another beautiful area. Starting over in Succoth in Arrochar, we travelled the machines and plough over a high track, far above the main road, to above the Tarbet Hotel. When we got there, we found the ground very steep in places and with the railway track below, it was too dangerous to work on most of the hillside.

44. Loch Long looking towards Succoth

Arrochar

DEEP DRAINING

Back in my day, there was a naval base at Arrochar. Unarmed torpedoes were fired down Loch Long from the naval base, and also from submarines. The torpedoes were then collected by a boat and returned for analysing. The base was closed in 1980. Arrochar had a buzz about it at that time, because it was also a favourite place for people from Glasgow to go hillwalking. The mountain up above Arrochar is called The Cobbler (real name Ben Arthur) and is just short of a Munro. It is in an area called The Arrochar Alps.

45. Glen Etive looking towards Beinn Fhionnlaidh, one of the areas we deep drained. Now most of the forest has been felled.

Glen Etive

DEEP DRAINING

Glen Etive was another fantastic area, until we spoiled it with the ploughing and deep draining. About a hundred years ago there were about eighty to a hundred people living there. Now there are only six to ten people living there all year round. Most of the people back then were farming, and working on the shooting estates. There was a ferry boat that came up Loch Etive with supplies and tourists to the pier, and then the tourists were picked up by small buses, taken to Ballachulish railway station and then back to Oban. My friend's father owned and also drove these buses, and still lives in Ballachulish. The two shooting estates are still there and have accommodation for guests.

My friend Walter, who was a shepherd on the Glen Coe and Glen Etive hills all his life, can remember looking down from Beinn Maol Chaluim into the corrie above the forest in Glen Etive and counted around about two hundred and eighty stags. It is changed days now, as deer are getting very scarce. You will still find some tame deer at the side of the road that the tourists film with their camera or phones, and the tourists also feed them.

When I worked there, the post office and school were still open, and a few families still lived there. There were some amazing people, including three men—Alex, Adam and Frank—who were stationed there after the Second World War to work in the forest. I think they were from Poland, because the locals called them The Poles.

Adam and Frank never left the glen the whole time they lived there, as far as I know. A local grocery van delivered food and drink once a week. On a rare occasion, Alex would go up to the Kingshouse Hotel. Adam and Frank never drank alcohol, but Alex did. I was living in a caravan in the farmyard where they lived and, occasionally, a few hours after the vodka delivery arrived, you might find Alex, a big man with a big moustache, lying anywhere. He had an altar in the wood which he used for his pleasure. He was into magic mushrooms, and the combination of the vodka and magic mushrooms left him out of the game. Adam was a very quiet man who kept to himself. Frank, on the other hand, was something else! He thought the war was still going on, that the local forester was a German spy, and German submarines were coming up Loch Etive. He also thought I was a British intelligence officer.

On Fridays I usually left the glen and went home to Strachur, and Frank would give me a letter written in Polish to give to my boss. One day, Frank said to me, 'Come and I will show you where I stay.' I had never been in their farmhouse. We went into the kitchen, and all that was there was a cooking stove, a few pots and so on, a table, and a few chairs. He then said, 'I will show you my room' (which was upstairs). Being young and not too sure what I might be getting myself into, I proceeded upstairs and into his room, which was very sparse and only had a bed in it with bed linen. He then went behind the bedroom door and produced a crowbar and said, 'Peter, this is for the Germans when they come. I will hit them over the head with the crowbar.' I left his bedroom, never to return!

Years later, the Forestry Commission sold the forest to the local estate, and Alex, Adam and Frank were moved out of the glen to Glen Duror, which is about thirty miles away. After being in Glen Etive for so long, they struggled; Frank took his own life in the forest, Adam found some relations in Poland and Alex, as far as I know, died in Duror.

Also living in the glen at that time were Jimmy and Edie, and their family. They were the kindest people you could ever meet. As I was staying in a caravan, I would come back after the weekends, and there would always be a bag full of cooked food hanging on the door handle.

Jimmy's past life was something else. He never spoke about it at the start of our friendship, until one day when I was introduced to a friend of his who had come up from London. Bert was a big tall blond man with some scars on his face; he wore tinted glasses. That night, we decided to go up to the Kingshouse Hotel for a drink. I was driving, as Bert did not have a car. After a few drinks, Jimmy and Bert decided to buy a carry-out of drink. Bert insisted on buying, and produced £100 out of a very large roll of notes for the carry-out. This happened regularly when Bert was in Glen Etive. After the introduction to Bert, I said to Jimmy, 'What does Bert do for a living?' Jimmy said, 'We call him the milkman.' On other occasions, some more of Jimmy's friends would arrive up from London in a Rolls Royce. I never met them.

I discovered one night in Jimmy's house, after a few drinks, that he was sort of a bodyguard for the rich and famous, and some of the underworld. One night, again after a few drinks, he showed me some photos: there he was, in all the photos with these famous people.

After I left Glen Etive, I heard Bert had died (I don't know how he died). Jimmy buried his ashes in the old sheep fank at the side of the road and marked it with a stone which just said 'Bert'. The stone is probably still there to this day.

There are so many stories from Glen Etive. This one stands out in my mind. We used to enjoy going up to Kingshouse Hotel for a drink, as there was not much to do in Glen Etive, especially if you were young. One night we decided to go, but my car had broken down. Jimmy being Jimmy said, 'Do not worry.' We went around to the back of his house and there was an old wreck of a Land Rover with no roof and no fuel tank. Jimmy said, 'I will get it going, as I use it to collect firewood.'

About half an hour later, with me sitting in the back (I don't even remember there ever being a seat in the back), holding a five-gallon fuel tank with a fuel hose coming out of it, we headed up the glen about twelve miles to the Kingshouse Hotel. We made it to the main A82 Glen Coe road, but decided to leave the old Land Rover in a lay-by on the Glen Etive road and walk the last mile.

After a few drinks, we headed back to the Land Rover. When we got there, we found a young man (with a big rucksack) looking for a lift down the glen. Jimmy being Jimmy said, 'Of course. Jump in the front.' We had no sooner started, with me holding the fuel pipe in the can, when Jimmy said to the young man, 'Can you sing?' and the young man said, 'No.' Jimmy shouted, 'You better start to sing, or I am going to fall asleep!' The young man sang all the way down the glen.

Jimmy was our banksman when we were deep draining, showing us where to go and helping if we had any problems. One day, we turned up a nest of weasels or stoats (we were not too sure) with the plough. Most of them were very small and very young. Jimmy said, 'What lovely wee animals', and he bent down to pick up one of the small animals when it grabbed his finger and hung on. He gave out a loud squeal, and had a terrible job getting the wee animal to let go of his finger. Jimmy and his wife Edie moved to the village after the forest was sold, and Jimmy is no longer with us.

Glen Etive was another place that was covered in the old Caledonian pine roots. We were ploughing below the road one day and we got completely bogged. The two tractors were bogged, and the deep draining plough was stuck below a Caledonian pine root. We had a problem, so the local forester called up our headquarters (Chapelhall in Glasgow—the Big Boss) and told him what the problem was. The next day the Big Boss arrived with about another six or seven forestry people, and he was going to give everyone a demonstration on how to unbog tractors. By about four o'clock in the afternoon, after breaking all the winch ropes, the tractors and plough were still stuck. The Big Boss did not do very well. We eventually persevered and unbogged them ourselves.

I met an old friend in the village a few days ago, and he reminded me of the night I probably saved him getting severe frostbite, or worse. I was heading up to the Kingshouse Hotel for a refreshment and ran into a severe snowstorm. The snow was probably about a foot deep on the road and visibility was very poor. Something caught my eye at the edge of the road. It was two men, Dickie and Drew, whom I knew from meeting previously in the hotel. They were like two snowmen and were so cold that it took a few minutes for them to speak. It was the late sixties—the Hippie Days. Dickie and Drew were only wearing denim jackets and jeans. There were no cars on the road that night, and if I had not come along it might have been a different outcome. The snowstorm was so bad that we had to spend the night in the hotel.

One Friday after work, we were sitting in the forestry van in Glenachulish, about to go home, when the forester appeared at the window. I rolled the window down and he said, 'The gamekeeper has shot a stag high up on the mountain outside the forestry fence, and I don't expect any of you to go up the hill and bring the stag down, but any of you are welcome to go up and get it for yourselves.' All the men said, 'No way are we going up there on a Friday night.' I phoned up a friend and told him the story, and as we were both very fit back in these days, we agreed to go up and get the stag.

After a long walk (no roads on that mountain at that time), we found the stag; it was a big one. After a quick inspection of the stag my friend said, 'It is not the best, but it is too good to leave on the mountain.' We fixed a rope to the stag to drag it down the mountain, and set off; it was starting to get dark and we did not want to draw attention by turning on our headtorches. We were getting close to the road, and my friend said, 'I will phone a friend to come with his car and trailer to pick up the stag and take it to Fort William.' Next day we got our reward: the stag had turned into pork (the stag was traded for two pig carcasses).

46. Ski Lodge, Bridge of Orchy

Bridge of Orchy

DEEP DRAINING

This was another outstanding beauty spot that was not very big at the time. Nowadays the forest stretches all the way down most of Glen Orchy. The forest had been planted, so we were only deep draining. Once again, we encountered the Caledonian pine roots. I cannot remember much about the work I did in this area, apart from the usual destruction and a good hotel (pub). Bridge of Orchy Hotel is a roadside hotel which caters for tourists and is one of the main hotels on the West Highland Way. In the winter, it also caters for some skiers when the snow conditions are good. It has been transformed from a small, quiet establishment to a busy hotel, popular with people walking the West Highland Way. Of all my time working in forestry, the old part of the forest at Bridge of Orchy was the only place that I ever saw a capercaillie.

There is one thing that stands out in my mind about Bridge of Orchy. It was back in my early days in Glen Coe, before the modern snow gritters had come on the scene. I cannot remember how I got involved, but I ended up on the back of a county council lorry, shovelling grit on to the A82 road, all the way from Ballachulish to Bridge of Orchy—about twenty-five miles. It was a cold winter's night, and there were four of us on the back of the lorry which had a small tin hut on the back for shelter. We all had shovels, and we were spreading the grit out the back of the lorry onto the road in absolute freezing conditions, especially over the highest point, just past the Glencoe Ski Centre. We had run out of grit a few miles before Bridge of Orchy, and I was expecting the lorry to turn around and go back to Ballachulish, but it kept going—all the way to the Bridge of Orchy Hotel. The lorry stopped, one of the men went into the hotel, and appeared out five minutes later with a half bottle of whisky (payment for keeping the road open). We all had a nice wee dram out of the half bottle (except the driver); it kept us warm until we got back to Ballachulish.

There was a time when we had a big snowstorm in Glen Coe, the A82 road was blocked for several days, and people had to be rescued from their vehicles, and from all the climbing huts. The police decided they would have to try and get up the glen with a snowcat (tracked machine) to get provisions to the remote houses. They also had to check the road, as there were so many cars stuck in the snow, and check all the climbing huts again.

My friend the police sergeant asked me if I would drive the snowcat, as my job involved driving tracked machines. We met at the police station and picked up the snowcat, which was on a trailer. We proceeded up the glen until we met the big snow drifts. We unloaded the snowcat, and just as we were about to leave to go up the glen, the police sergeant said, 'I will drive,' which startled me for a minute, as I thought he did not know how to drive the snowcat. Then he told me that, the day before, he had taken the snowcat up a glen with hay for the farmer, so he learned how to drive it.

Off we went up the glen and we ran into some very deep drifts of snow up on one of the high points of the road. What we did not realise at that time was that there were cars underneath us as we were travelling. We kept checking the climbers' huts and vehicles as we went all the way up to the Kingshouse Hotel and the ski centre, still with the police sergeant driving the snowcat. He was like a wee boy with a new toy. The police sergeant made a few phone calls and it was then that we discovered that we had travelled over vehicles buried under the snow (the owners had been rescued the night before). Then he drove his new toy (snowcat) all the way back down the glen. After that experience I knew what to buy him for his birthday.

The Glen Coe road was blocked for a day or so while the snowplough and diggers cleared the snow away from the buried cars up on the high point of the road. I think some of the cars got a bit damaged clearing the snow from around them.

It was around that time that the Kingshouse Hotel was blocked in and as I was working at the ski centre, I took the snow groomer down to the hotel, cleared the road and also cleared the car park, so that vehicles that were stuck could get out. That was probably the biggest snowfall I had seen in my time up at the ski centre.

Before I lived in Glen Coe there was an even bigger snowfall, with lots of vehicles stuck for almost a week, and some of the people were put up in the Kingshouse Hotel and the Bridge of Orchy Hotel. There were photos of that big snowfall in the Bridge of Orchy Hotel.

47. Glen Orchy road end looking towards Dalmally

Dalmally

DEEP DRAINING

Dalmally was another fair-sized forest, stretching from the road all the way up to Succoth Glen and part way on both sides of the road almost as far as Tyndrum. At the top of Succoth Glen, we came across an old settlement. I remember there were quite a few old ruins, out near the top of the glen. When we were ploughing around the ruins, the plough turned up several clay smoking pipes which I gave to Glencoe Museum. There was also part of an old tartan kilt, which fell apart, a round bottom glass bottle, and some carved wood. Lower down the glen, we ploughed up some charcoal pits which again were quite common, with the iron foundry being quite close at Taynuilt. We also deep drained part of the hillside on the road to Tyndrum, below Ben Lui.

When I was in the Glencoe Mountain Rescue Team, we were called out to the crash site of a fighter plane on Ben Lui. We were transported by helicopter to the top of Ben Lui to look for the plane, and landed somewhere near the summit, in the mist. We searched for some time and then heard on the rescue team radio that the plane had been found and traces of blood were found somewhere at the crash site. As we were about to head down the mountain, another call came over the Mountain Rescue radio to say the search for the pilot was still on, as the blood that was found at the crash site was from a rabbit or hare.

A while later while we were still searching for the pilot, we heard on the radio that he had been found and, unfortunately, had been killed. He had managed to eject himself from the plane, but was not alive when the rescuers found him. When we were leaving our search area, the mist was very bad and the compass we were using was not working, so our group became a bit disorientated. The mist started to lift, and I spotted the hill I knew very well, as I had deep drained that area.

While I was working in Dalmally, the forester found accommodation for me with an old lady who lived in an old two-storey house close to the River Orchy. After the first day of working in Dalmally, I went to the house, knocked on the door and introduced myself. The old lady was dressed completely in a long black dress, had long white hair and was well over retirement age; she was also very polite and very well spoken. She took me into the living room and introduced me to Angie, another man who was lodging there. I knew him; he also worked in the forestry. We then proceeded upstairs, followed by her wee white dog, and she showed me my room.

After having a wash, I went downstairs for my evening meal where Angie was sitting; he told me a bit more about the old lady. He said she was a lovely old lady, maybe a wee bit eccentric, was a great cook and quite often did some strange things in the night.

After a lovely meal, I headed to the local pub, as there was not much else to do in the house and Angie liked to read. After a good night in the pub, I walked back to the house where I was staying. Everybody seemed to be in bed, so I made my way upstairs and went to bed. I was woken by a chop, chop, chop noise and being half a sleep, I looked at the time: it was two o'clock in the morning. I went over to the window to where the noise was coming from, and pulled back the curtains to see the old lady, with her dog, chopping firewood with a big axe. At that time and being young, I remembered watching scary movies, so I thought that I had better be careful here, so I jammed a chair up against the door handle to stop anyone entering the room.

Next morning at breakfast I told Angie of my experience, and he said, 'Don't worry. These things happen all the time.'

Just before the Falklands War, I worked on another BBC film made by Hamish MacInnes up on Ben Nevis. Four climbers were going to climb two routes up the face onto the summit of Ben Nevis, and a French skier was going to ski straight down off the summit. There was a squad of Royal Marines helping with the making of the film, and one of the marines was part of the climbing team, just before they went to the Falklands.

The marines were camped out on the summit for most of the film, looking after all the filming equipment. The weather was terrible, as it often is on Ben Nevis; it proved to be quite a difficult film to make.

We also had another location, at the bottom of the climb near the CIC hut. Most of our job was working with the helicopters, getting all the supplies and film equipment to the two locations.

One day, the helicopter came in with a load which was hanging from a long metal rope, and I went to release the load from the helicopter, as that was my job, when a sheet of insulation flew up in the air towards the helicopter (the insulation was used to insulate the tents to stop all the filming equipment freezing up). Davy the pilot spotted it coming straight at him; he had a quick look below him and thought I was clear. He released the load from the emergency hook directly below the helicopter, to stop the insulation from hitting his rotors and crashing the helicopter. He then shot off. The load landed beside me, but the metal shackle and lifting gear hit me on the top of my head. Back then, unless we were climbing, we did not wear safety hats—just the normal woolly hats. Davy saw what had happened, landed the helicopter and shot over to where I was sitting, with blood all over my head and face. He said, 'Come on Peter, get in the helicopter and we will fly down to the hospital', which we did but, with no landing area at the hospital, we landed a bit away and got a lift to the hospital.

When we walked into the Emergency Department, there was a queue at the reception and with my head and face covered in blood, we jumped the queue and went straight to reception. The girl behind the desk said, 'Can you not see there are people in front of you! Get to the back of the queue.' That was us put in our place. After getting patched up, we got a lift back to the helicopter and headed up the mountain.

One day, the mist was down very low, so the helicopter could only fly halfway up Ben Nevis (which was often), and the film crew needed some equipment taken to the summit. It was the job of all the climbers who were helping on the film to get filming equipment to certain locations on the mountain. Davy, the helicopter pilot, managed to get us above halfway and as he could not land, he rested one of his helicopter skis on a large cairn and we all managed to get out, along with the BBC filming equipment. We had all the equipment in rucksacks for carrying up the mountain, so off we went, heading for the summit of Ben Nevis. We hadn't travelled very far when we had to stop to put crampons on our boots, as it was getting very icy and there was a big drop to our left down into the corrie below. Ronnie, one of the climbers helping, took his rucksack off to put his crampons on and when he turned around to get his rucksack it was nowhere to be seen. The rucksack had slid off the mountain, down into the corrie below.

Later we discovered what was in the rucksack: a very expensive part which was needed for filming; someone said it was worth £20,000 or so. The BBC managed to get the part replaced, and a day or so later Ronnie was working up in the same area. The weather was a bit better, so he decided to have a look over the edge, spotted the rucksack caught on a rock, and eventually returned it to the BBC.

Ronnie was getting married quite soon after we had made the film, so the producer Mike got old broken parts from a camera and put them in a frame and that was Ronnie's wedding present. Ronnie has the frame hanging on his wall in his house.

Another day, the helicopter had a problem and could not fly, and we were needing fuel for the generators at the bottom camp. We got carrying frames (from somewhere) for carrying five-gallon cans. Ronnie, Davy (the pilot), the Royal Marine sergeant, some Royal Marines, and I set off from the forestry road, up to the bottom camp. It ended up being a race to see who would get there first with the fuel cans. The first man up was Davy the pilot, followed by me, Ronnie and the marine sergeant (we were all together), with the other marines coming behind. Davy was extremely fit, as he proved that day on the mountain. Eventually, everybody managed to complete all the filming.

48. Chris Bonington and myself.
We were working with Hamish MacInnes on a film about Chris Bonington on Aonach Mòr, Fort William

49. Glen Orchy (from the bridge)

Glen Orchy

DEEP DRAINING

Glen Orchy is another stunning area, with the River Orchy running beside the road. This area had already been ploughed, and we were sent to deep drain it. This was another glen with old settlements, some high up in the glen. Lower down, close to the river, was an old ruined farmhouse with outbuildings. Today there are only about two farms and four houses in the whole area.

One morning, high up in the glen while I was working on the plough, my hand got jammed in the back of the adjusting bolts, which severely squashed my thumb. Archie and Donald, my workmates, had to dismantle the plough where my thumb was jammed. When we managed to free my thumb, there was blood everywhere. It was a long walk back to the road. We went to the doctor in Dalmally to find he was out somewhere else, and his wife told us to go straight to the hospital in Oban. At that time, there were holdups on the

Oban road at the Pass of Brander. She said, 'I will phone ahead to the road works, tell them it is an emergency, and to let you straight through,' which they did. Archie drove me to Oban hospital, where I spent the night ... with the top of my thumb missing.

Most times we crossed the River Orchy in a Land Rover, a few miles from the bridge. Crossing the river by driving through it was fine most of the time, except after heavy rain, when the river was in spate. At other times, when the river was low, we would put waders on and cross over on foot. On one occasion, the machine was broken down on the other side of the river. With the river being quite high, we attempted to cross with the Land Rover, but we underestimated the force of the river. We were swept about fifty yards down the river. My workmate Archie was screaming his head off at me. Fortunately, the wheels of the Land Rover managed to get a bit of traction on a gravel bank and we were able to get out of the river and up to the bank where we had started. When we got out of the river, the height of the water on the side of the Land Rover was up over the door handle.

On another occasion in the same area, one of the machines broke down and the mechanic, Jock Martin, came out from the workshop in Barcaldine to fix it. We crossed the river in the Land Rover and then started to walk up the hill. I was carrying Jock's tool bag. He had a walking stick, as he was getting on in age, but he was still a hardy man. After about ten minutes of

50. Glen Orchy, where we used to cross the river in the Land Rover

walking, a squeaking noise started. Jock shouted, 'Gimme the tool bag', and he proceeded to take out an oil can. He rolled up his trouser leg to expose a wooden leg and started to oil the joint on his wooden leg. Jock was one of the toughest men I ever met.

Quite often when we were travelling across steep ground with the tractors and plough, we would slide down the mountain and, on occasion, one or two of the tracks would come off the machine. Sometimes the tracks were a way up the mountain behind us, and as we had winches with a thick wire rope, the only way to get the tracks down was with the winch. The process of getting the tracks back on the machine was a major job.

One day, while we were draining that area, the forester was sitting down on the furrows, working out how much to pay us—or not to pay us as was often the case—when he noticed an old musket ball on top of a furrow, and put it in his pocket (he did show the musket ball to us). I do not know what happened to it. The area where he found the musket ball was in front of an old farm which was still standing, but turning into a ruin.

One night in the local pub, an old man from the area said that seven hundred men came out of Glen Orchy (I think he must have included Dalmally as well) to fight in Culloden.

Glen Orchy has one of the most beautiful rivers in Scotland; it runs the full length of the glen and is very popular with campers. Halfway down the glen there is a metal bridge at one of the most beautiful parts of the river. I was working up on the hill, on the far side of the bridge. Back when I was working in Glen Orchy, the bridge was wooden, with gaps in between the wooden decking. You could look down through the gaps and see the river below.

One day, I had problems with my brakes on the machine, so I took it down to the bridge so I could work below the machine on a dry surface. There were a few tents on the bank just above the bridge, very common back then. Some of the campers were moving around, as they do first thing in the morning. I proceeded to crawl under the machine and started to unbolt the metal plate which had four bolts holding it. I managed to undo three of the bolts, and was working on the last bolt when something caught my eye through the gaps on the bridge. There below me was a naked woman, walking down to the river to wash herself. Being a young man, I was distracted, dropped the last bolt into the river, and the metal plate landed on my chest. Over that bridge and high up the mountain is one of the old settlements, and where I lost the top of my thumb.

51. Looking towards Glendaruel from the Tighnabruaich Road Viewpoint

Glendaruel

ROAD WORKING AND HILL PLOUGHING

Glendaruel is an area I knew very well, as I worked on the new road there and lived just over the hill. I worked with a lot of Irish men back then, which was very common on construction sites. Most of the Irish men who were not married were staying in a camp down the road from the hotel. On pay nights, most of the Irish men who were not married would go to the Glendaruel Hotel (sadly it is closed now), and some would hand over their pay packet to the barman, saying, 'Put my name on it and throw me out when it's empty.' There were no credit cards back then. The pay packet rarely lasted the week.

When I was working on the Glendaruel road, we used to do all sorts of work—no Health and Safety back then. We were needing stone for dyking and I was the one who went to blast some stone out of the quarry. The quarry was above the road, so two men had to stop the traffic. As we wanted good stone for the walls, I did not use too much of the explosive. I lit the fuse and retreated (back then we used fuses). With all the traffic waiting on the road, and me waiting for the big bang, nothing happened. After about fifteen minutes, with car horns blaring, I nervously edged my way back up to the quarry to discover that the fuse had gone out.

Back then, along each edge of new roads there was a concrete haunch to stop the tar from breaking away. As we were working there, the boss asked my mate Mike and me if we would take over the contract, as the previous contractors were leaving to go to Australia. The system they were using was shovelling the concrete out of a side-tipping lorry into the track at the side of the road. This was very slow and hard work.

The boss supplied the concrete and the lorry, so we invented a new system, as there were no ready-mix concrete lorries back then. We divided the lorry in half, running at an angle with wood so at the back of the lorry the space for the concrete was wide and then narrowed down to about a foot at where the concrete was going to come out. We then hung a sheet of corrugated tin below the door at an angle to form a chute.

It was a great success, and all it took to work was one man on the lorry, lifting the tipper, and one man, with a rake, at the back of the lorry to adjust the flow of concrete as it was flowing out. The concrete was already mixed at the batching plant. We were covering a lot of ground with the new system, and as we were getting get paid by the distance, we were rubbing our hands together, thinking of the big pay packet when the boss appeared and laughed at our invention. He said, 'Well done, boys. That is a great invention, but I can't pay you all that money for the distance you have covered.' As it was his concrete and his lorry, we came to an agreement and carried on working. Nowadays the ready-mix lorries do the same job.

Another time, we were building a bridge for the new road, and were putting in the steel for the foundation. We had erected a wall with sandbags around the steel to keep the river out. Without warning a big wall of water came towards us and knocked down the sandbags, trapping one man as his boots got caught in the steel. We had to go under the water to free the man. Thankfully he was alright. Afterwards we discovered the reason the river had flooded: there was some sort of dam higher up the river, and someone had released the water without telling us.

Another story that comes to mind is while working on the Glendaruel road. One day after heavy rain one of the new culverts on the road got blocked, and the water was coming over the road. Some of the wood shuttering had gone down the pipe and jammed there, forming a large dam behind which the water and hillside gravel got stuck. There were discussions on how to clear the culvert, and the foreman said that someone would have to crawl up the pipe and free the stuck wood. As one can imagine, there were no volunteers, as the culvert was quite long, so the foreman said, 'I will go, and take a small metal bar with me, and free the jammed wood.' Off the foreman went, up the culvert with his metal bar to where the wood was jammed, with some of us watching up the culvert to see what would happen. We all guessed what would happen, and it did. The foreman levered the stuck wood free, and we disappeared away from the bottom end of the culvert, followed by the foreman, wood rocks, and lots of water. The foreman come out at such a speed that he landed about twenty yards down the hill and, amazingly, he escaped without a scratch on him (lucky, foolish foreman).

I was hill ploughing on the mountain behind the quarry. Back then we had no roof (cab) on the tractors. I was ploughing up on the crest of the mountain and I had a banksman—a spare man for safety—when a big thunderstorm appeared right above us. With the tractor being all metal, I thought it would be a good idea to get off the tractor.

The banksman and I were sheltering on each side of the bogey wheels (a wooden trailer with tractor tyres and made of wood for taking the fuel out to where we were working) when there was an almighty bang and flash. The lighting struck in between us, hitting the wheel stud, shaking the ground all round us and leaving a great big burnt mark on the wheel stud. One can imagine the speed we disappeared off that mountain.

A similar thing happened years later, when I was building fences with my workmate Neil on the crest of a mountain in Glen Coe. We were fencing around the forest on top of the mountain ridge when a thunderstorm struck. Lighting was flashing all round us, and we had to pass through the fence we were building to get down the mountain. We just got through the fence when the fence wire lit up all the way behind us, sending shivers down our spine.

Nowadays, when you drive through Glendaruel you see mostly forest trees but, working there in my day, it was mostly farms.

It was about New Year in Glen Coe when a few of the mountain rescue boys appeared at the house, as you do at New Year. After a wee while looking out at the river, which was very high, someone noticed I had an old dinghy lying in the garden. Someone else suggested that we launch the boat and have a bit of fun. The plan was for two people to go into the boat, with the rest of the boys running along the river bank to the green pool (a big pool which was normally quite calm), then to throw a climbing rope to the boat, and pull it ashore before we reached rough water and a gorge.

We drew lots to see who would go in the boat; Doug and I were selected. We launched the boat, got in and off we went—with no oars—at high speed. We travelled so fast that the boys on the bank could not keep up with us. The boat, with us in it, shot through the green pool and we headed for the gorge, holding onto the side of the boat, as we had no oars. By pure luck, there was an overhanging tree in front of us, and we both managed grab it, swing up onto it, and slide along the tree onto the bank. We took the boat rope with us and managed to save the boat.

52. Looking towards Inverkip Forest from Dunoon

Inverkip

HILL PLOUGHING

Inverkip was just across the water from Dunoon. It was not a very big area back then. Two things stand out in my mind. Firstly, I was lodging with my relations in Gourock, and I used to eat most nights in a Chinese restaurant in Gourock. The food was very good, and the Chinese owner got to know me quite well. One night when I was in for a meal, he said, 'Peter, you work in forestry. You get lots of foxes. Foxes very good for curry.' I never went back to that restaurant.

The other occasion was a Friday night, and I had just been paid. I went to the local pub down by the lighthouse, I had a great night as usual, and went back to where I was staying. Suddenly, when I was about to go to bed, I realised that my wallet was missing—with all my pay in it. I got dressed and shot down to the pub. The staff were cleaning up, as the pub was closed. I banged on the door, one of the staff opened the door, and I told him my problem. He asked my name and I told him. He appeared back with my pay packet, with the wallet unopened. That made my night! Thanks to the staff back then!

The work we were doing—hill ploughing and deep draining—ran out, and new modern machines took over, so our deep draining and ploughing days were over. We were offered jobs back in the forest where we lived. The forestry jobs varied: felling trees, timber extraction, building fences around the forest, and general forestry work.

I started off in the felling squad, cutting down the trees that had been planted around forty or fifty years before, as this forest was planted just after the Second World War.

I quite enjoyed cutting down the trees, cutting the branches off and cutting the tree into the size for the sawmill. It was also a dangerous job working with a power saw, with bad ground conditions under our feet and, as usual with the forestry, when we worked hard and made good money, the forester would cut our price. It seemed to be a thing back in these days that if you worked hard to earn good money, you got penalised by the forester cutting the price.

After felling the trees, I went into timber extraction. This consisted of a tractor with a winch and tower. You had to set up the tractor and winch at the start correctly; if not, you were wasting your time with the skyline: the metal rope which ran all the way up the mountain and shackled on to a tree at the top. The haul-out and the haul-in ropes also had to be set up properly. A carriage was attached to your haul-out and haul-in ropes and then hung over your skyline. It was then sent up the mountain to the man in the wood who would radio down to stop where he wanted it. He would then attach the logs to rope slings, which he had fixed on the logs, then onto the hooks on the wire rope, and then call the man on the tractor to winch in. Sometimes you could have as many as eight logs coming down the mountain at the same time. At the bottom, you had a safety device fitted on the skyline. As the logs were coming towards the tractor at the bottom, the carriage would run over the safety device and the logs would drop down onto the log pile. It could be dangerous at times, as the logs, when dropped, could slide towards the tractor. The other dangerous problem we had was that we were dragging around eight logs at a time (depending on the size) and the log pile at the bottom would get very big and sometimes unstable. The job for the man at the bottom was to climb up onto the pile of logs and release the rope slings.

I was at the bottom one day, and the pile of logs was getting very high. I always took a metal bar with me to lift the logs in order to release the rope slings. On the way down the log pile, the whole pile started to move, and my welly boot (with my foot in it) got trapped in the moving logs. By pure luck, I managed to stick the metal bar in the log pile just above my welly boot, and all the moving logs piled up against the metal bar. We always carried a radio on us, so I radioed my workmate Neil and told him to come down straight away, as I had a serious problem. Neil managed to get another metal bar and lever the log that was jamming my foot enough to get my foot out of the welly and then pull the welly out. That is what I call a close shave!

After we reported the incident to the forester, they decided to get us another tractor with a log grab on it so that we could lower the pile of logs before it became too high.

When all the logs were dragged off the area we were working on, we had to move the tractor and winch along the road to the next area. The main skyline was attached to a metal drum on the rear wheel of the tractor. In order to haul the skyline in we had to jack up the tractor and spin the wheel so that metal rope would wind onto to the metal drum. As Neil was up the hill dismantling some of the other equipment, I was on my own, winding in the skyline on the spinning tractor wheel. I had my metal bar guiding the skyline onto the metal drum when the metal got caught on the spinning wheel and came around at a terrifying speed. It hit me above my nose, in between my eyes. I was knocked out—stone cold—for some time, as when I eventually came round, all the skyline was on the drum and the wheel was still spinning. For years after that incident I could not smell, until a few years ago, when I had a heart attack. I had two stents fitted and my sense of smell came back.

After the Forestry

I was offered a job with a house back at Strathlachlan, building houses for my old boss who lived at the castle, so I left the Forestry Commission. It was going to be a big development, and I was involved in building the first three houses and groundworks.

One day, my first wife Rosemary got a phone call to say that her dad Allen was ill and had had a stroke. We had to go back to Glen Coe, as there was a croft and animals to look after.

The day before I was leaving to go back to Glen Coe, my boss George and I decided to have a farewell party on the building site. There were only going to be three or four people, but the word got out that there was a farewell party in the building site, so we had a few more guests than planned. Being a Friday, the party went very well, with a lot of wee drams flowing.

The main party area was around a manhole that had not been used at that time. As the party flowed, George—my boss—slipped down into the unused manhole, with his bottom on the base and his legs and top part of his body sticking out each side of the manhole. We were laughing so much that George was neglected for a while. Eventually we pulled him out.

I became a part-time crofter, looking after cows and working as well. It was a lot of work, especially at hay time, as we did not have a tractor back then. There was always something happening on the croft; one day I had a bullock that was not very well, so I phoned the vet in Oban. I explained what was wrong and he said he was too busy to come up, but he would send up treatment on the Oban bus for me.

I got the package from the bus and headed for the byre where the sick bullock was lying on the floor. The byre door was two half-doors, so I closed the bottom door and then read the instructions from the vet then proceeded to take out a syringe with a long needle. The instructions said to fill the syringe from bottle and stick the needle in the back end of the bullock, which I did. The bullock jumped to its feet, knocked me over and then jumped over the bottom of the byre door. The bullock spent the next two days in the field with the needle sticking out of its rear end before I finally managed to remove it. We built a house on the croft. Later, we extended it to become a guest house which we ran for many years.

Peter Weir

I also went back to work on construction, mostly driving excavators for a while. I was working in Oban on an excavator for quite some time digging up parts of the pavements to put in new cables. Almost every few yards there would be power cables and water pipes, so quite often there would be a flash or a spout of water. One day, while digging in the centre of Oban, I pulled up a black cable which looked as if there was a sort of cover on the end of it. One man was behind me in the new track. He proceeded to poke the end of the cable with his shovel, there was a big flash, and the man was thrown back along the new track for a few yards, landing on his back in the track. Luckily, he was wearing wellies; that probably saved him from being electrocuted.

After Oban, I worked on building the old Glencoe Visitor Centre, driving a 360-degree digger, mostly building the roads and parking spaces. I remember at the far end of the road there was an old ruined house which my friend Walter, who was a shepherd in that area, used as a holding pen to twin lambs. I was ordered to knock the old house down, as it was in the way of the development. I was also in the river with the digger, loading lorries with gravel for the new roads and car park. While I was working at the old visitor centre, I had an interview for the manager's job at the Glencoe Ski Centre, which I got, and the rest is history!

One day, I was working at the old visitor centre with Innes, a local man who was a bit of a character. He said, 'I am going for a walkabout.' I was thinking he was going to the pub, which was across the river, as he did during some lunch breaks. Innes disappeared for years. Apparently, he hitched a lift with a lorry, went abroad and joined the French Foreign Legion. Some locals believed that story, and some did not. Whatever way you look at Innes's story, he was amazing character and was famous in the area for his pet dogs. I think they were Siberian Huskies—maybe a mix with wolf(!)—beautiful dogs. He had two and when one died, he replaced it with another one. Everywhere Innes went you would see one of the dog's heads sticking out his car window, especially at the coffee shop car park. Sadly, Innes and his dogs are no longer with us.

I lost my first wife Rosemary at that time to cancer. My daughter Rhona was only around twelve or thirteen years old and was struggling to go to school. The old Carnoch Restaurant came up for sale, and I bought it with a business partner, which hardly ever works, as happened in this case. After a while, I bought him out, and we went our separate ways. I ended up running the restaurant, with my daughter also working there.

After about two years, I was out on the road taking my OPEN sign in when a car stopped, the lady rolled down her window and said, 'Are we too late to get some food?' and I said, 'No. Go in and the girls will cook you something.' The lady and her son went in and the girls cooked a meal for them. The lady and her son were American, she was very attractive, and I took a wee shine to her, as they say. Paulette and I ended up getting married and are still married to this day.

Paulette's history is quite amazing. She is from Albuquerque, New Mexico and she can trace her family back to the late 1500s, when the Spanish conquistadors came up through Mexico to Albuquerque. There is a mural on a restaurant wall in the square in the centre of old town Albuquerque, and Paulette's ancestors are on that mural, showing the arrival of the soldiers in New Mexico.

Paulette has another string to her bow. Her ancestors used to look after Billy the Kid before—and when—he was on the run. This is documented in an old newspaper and a book. The ancestor was a Mrs Dow, and she stated that Billy was a nice young man and would come and spend the night in their house and play with the Dow children when he was on the run. The Dows had a store near Lincoln. This was where the cowboy war between the ranchers started in 1878, and lasted to 1881. Most of the ranchers were Scottish, English and Irish. Billy the Kid's boss was shot by the sheriff and some ranchers over a dispute in a place in New Mexico called Glencoe (what a coincidence). I don't think there is any connection to Glencoe where we live.

Pat Garret was a sheriff back then (not in Lincoln) and a friend of the Kid and was supposed to have shot and killed Billy in a place called Fort Sumner, New Mexico. Billy the Kid's grave is there, but some people say the grave is empty; others say there is a young Mexican man in the grave.

53. Mural of my wife's ancestors on a restaurant wall in Old Albuquerque, New Mexico

Pioneer Woman Recalls Days of "Billy the Kid"

NOVEMBER 19, 1933

Few people know that in Albuquerque lives a woman who intimately knew Billy the Kid. At her home on Marble Avenue, Mrs. Isabel Dow gave for the Albuquerque Record a vivid picture of the famous boy outlaw and the dangerous days in which he lived. Seated there in a modern livingroom, she carried us back 50 years to the Lincoln County War, one of the most notorious feuds in the history of this country.

Menacing Indian raids. Dark characters slipping in and out of the Dow store at Ruidoso! Every thrill of those untamed days! Guns were always ready for quick action. Often five or six of them were kept high on a rack where the children could not reach them. Mrs. Dow's father always slept with a loaded gun by his bed and a pistol under his pillow. Mrs. Dow helped to make bullets and load guns. She would first put in the powder, then a piece of buckskin and ram down the bullet till it would stick.

What a picture of the life of those days is revealed by so simple an action!

Mrs. Dow's parents lived at Antelope Springs in Estancia Valley and the 175 mile ride back and forth from Lincoln County took five days of tedious travel over trails in a Covered wagon. Often they would not see a soul or a house for several days. Mrs. Dow told me they always camped a mile from the road, had their fire beforedark and buried the embers in haste before the Apaches might see them in the darkness. She said that when she lived at Lincoln she would stand in the doorway of their house while men were shot to death in the street before her, and her frightened sister would beg her to close the door. But as anyone understands who knows this pioneer woman, Mrs. Dow was never afraid.

It was at Ruidoso where she best knew Billy the Kid. Mr. Dow and he were friends. Frequently the young outlaw came to their house and ate dinner and afterward snatched a few hours sleep on the lounge while they watched the trail for his enemies.

Mrs. Dow said that he always kept his spurs on and the two fancy pistols in his belt while he slept. She noticed that even when asleep one hand remained on a pistol. His horse was kept in the corral at the rear of the house, saddled and ready for action.

Some writers picture Billy the Kid as a cold blooded bandit killing innocent people, but Mrs. Dow said the real Billy the Kid was a clean boy, nice mannered and very jolly, that he was forced to his life by his crooked undercircumstances, and that nearly everyone sympathized with him. He liked children and would sing songs of the range to Mrs. Dow's young boy and girl, takethem on his back, and put them on his saddle.

She declared indignantly that the well known picture which is accepted every where as the famous outlaw, was someone else and not the real Billy the Kid, or Billy Bonney, as they called him in those days. She said that in all the time she saw him not once did he havelong hair. "I would know him anywhere—now, if he came in!" she said.

Albuquerque should be proud of this courageous pioneer woman who also happens to be the grandmother ofone of our high school students, Wilma Pine.

54. 1933 Newspaper article featuring interview with Mrs Dow

55. Billy the Kid's grave in Fort Summer, New Mexico

56. Graves of Billy the Kid and his companions

Another story about Paulette's ancestors was about two men who were attacked by the Indians in New Mexico up in the hills near Chilili, above Albuquerque. One was a postman; a gunfight had started, and the two men managed to hold off the Indians (Native Americans) through the day. The story goes that when darkness fell, the Indian squaws crept up on the men and scalped them. Indian men never attacked at night, apparently. That incident is also recorded.

Another ancestor, Robert McAfee, who had emigrated from Ireland in 1844, was traveling by wagon from Chilili, New Mexico to Albuquerque when they were surrounded by Indians. They were poised to attack; however, the Indian chief recognised Mr McAfee as the man who had once saved his life, so the Indians accompanied the McAfee family safely to the outskirts of Albuquerque.

I have visited New Mexico many times. During each visit, I have learned more about the history of the state, from when only the Native Americans lived there, to how it was claimed by Mexico and Spain. New Mexico was a territory until it became a US state in 1912. The Pueblo Indians often still live in the pueblos. There are 23 Indian tribes in New Mexico: Pueblo Indians (19 tribes), Apache (3 tribes) and the Navajo Nation. Paulette has a collection of arrowheads which her family found on their ranch in the Jemez Mountains.

Our Restaurant Time

Our old restaurant was getting a bit old and in need of major repair and so we decided to build a new restaurant with rooms, and a place for ourselves to stay.

After a lot of hassle with the local authority and all the conditions they put on us, the new building ran well over budget. Because of the proximity to the river, there were 18 lorryloads of concrete, and seven tonnes of steel went into the foundation; that was just to get to ground level. The local planning authority insisted we put in a new sewage system, which cost a lot of money and put us over budget. The new restaurant, when full, was only catering for around thirty people at a time. There were also two rooms which we let out, and our own accommodation.

Around that time, the visitor centre moved down to where the forestry campsite was, and this wiped out most of our daytime trade. We persevered, mostly doing evening meals, as the visitor centre was closed at night at that time. My wife did all the cooking, as she is an excellent chef, and I did my Basil Fawlty head waiter, but with the bank breathing down our neck we struggled. We decided to put the restaurant on the market and we thought it would sell no problem as a business, as it is in a stunning location. We were wrong, because everybody who looked at it decided not to buy it because of the competition just up the road. We decided to get a change of use from a business to two houses, did the alterations required and put it on the market. We sold the restaurant part to a lovely couple who are still there to this day and we continue to live in the other part.

One evening, during the restaurant days, a young couple came in, and I took them to a table which had a nice view overlooking the river and mountains. We had candles on all the tables, and I would light the candles as we went along. After lighting the candle, I proceeded to the kitchen to get what we put on the table for all our customers: a basket of bread. I took out their bread, took their food order and walked away. Suddenly, I heard a loud scream. I had sat the breadbasket on top of the candle and the paper napkin inside of the basket was on fire. I managed to grab the burning breadbasket, take it outside and put the fire out. That night we had quite a lot of customers, and as I walked in carrying the burnt breadbasket, I shouted out, 'Anybody for toast?' That caused a good laugh with the customers.

Peter Weir

After we sold half the property, we decided to buy a house out in Colorado in the USA, to use as a holiday home. The house was up in the hills above Denver at 8,500 feet. I loved the house and surrounding area—mostly forests with some large open ranches—cowboy country. At that time, we did a bit of skiing and a bit higher up the mountain were all the famous Colorado ski resorts. We managed to ski, but spent a lot of time working at the house and cutting down trees, as they were too close to the house because of fire danger. With the house being at 8,500 feet we had a borehole—well—for water as had everybody else living in that area.

After a few years, we decided to let out the house on a long let, to give us an income, as we were not getting to use it as much as we would like. We advertised the house and rented it out to (what appeared to be) a very nice woman who had two grown-up children. The let turned out to be a disaster, and as we were back in Glencoe (Scotland), we could not keep our eye on the house.

One day we got a phone call from Colorado from the lady who was renting the house to say the well had run dry and they had no water. We phoned up a company who dealt in wells to go and check it out, which they did. They said we would need to drill a new bore hole. $25,000 later (that is what it cost to drill a new well) the lady who was renting the house had water. Not long after that we got a phone call from a neighbour in Colorado to say that the people who were renting did a moonlight flit and the house seemed empty.

My wife's daughter, who lived in Colorado, drove up the mountain to check out the house. She phoned us and gave us the bad news: the house was *wrecked* inside. The tenants had been growing cannabis all over the house, as we had plenty windows for light; and growing so much cannabis was the reason for the well running dry. We had to fly out to Denver to repair the house, as all the carpets were destroyed, along with all the paint work, walls, doors, most of the woodwork, ceilings and so on. They had left all their old unwanted furniture behind. We rented a very large skip, and after about two or three weeks of repair work, including new carpets, we put the house on the market. It sold some time later. Ouch!

CONCLUSION

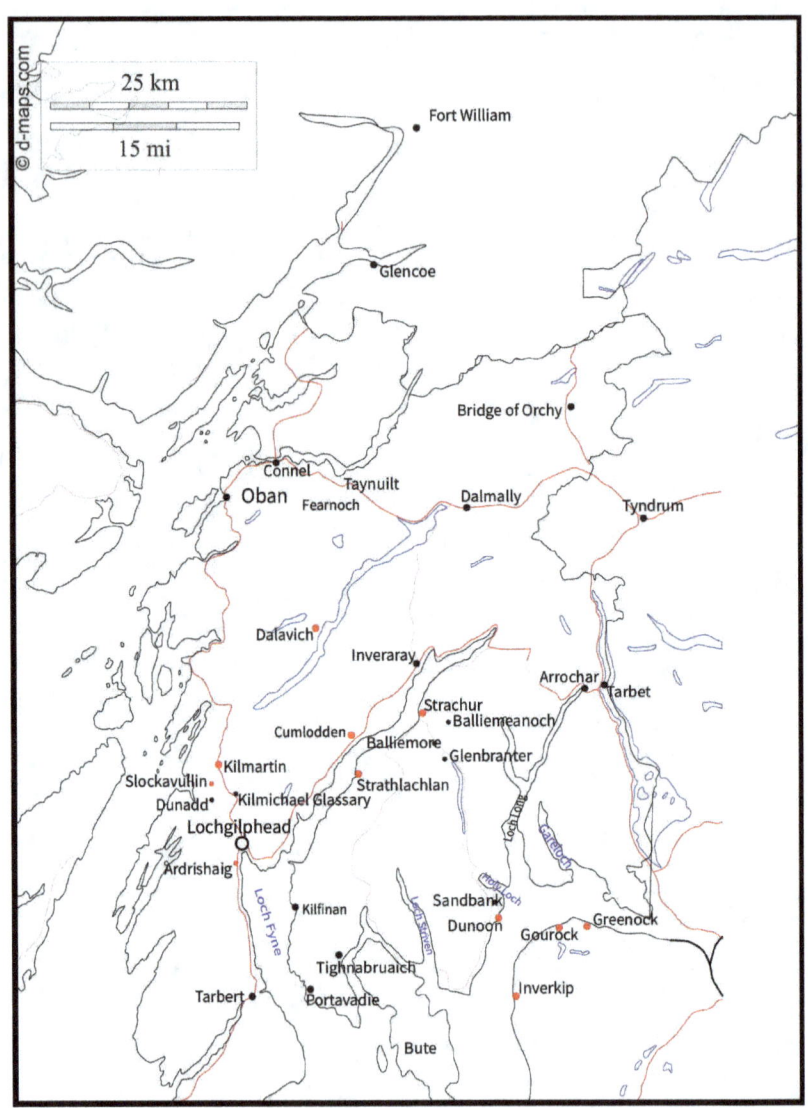

57. Some ancestral places, marked ●

Kilmartin

Kilmartin is famous for Dunadd Hillfort where Pictish kings were crowned. The Pictish kings might be some of my relations, as a lot of my ancestors came from that area!

58. Standing Stones at Kilmartin,
just down the road from where my grandmother was born

59. The foot indentation where the Pictish kings were crowned.
My foot fits perfectly!

60. Glen Etive area, where one of the seven hydro schemes will be built

In Closing

Nowadays we have the local coffee shop, the political headquarters of the Highlands, where the world is put to right nearly every morning, sitting around the round table with all the customers looking on in amazement at all the arguing.

I am all for green energy if is done right, and not just a free-for-all. I have explained in this book all the damage we did with the hill ploughing and deep draining to get a lot of the forests you see in Argyll today.

Here we are again, saving the world with hydro schemes. Once again, if they are constructed discreetly in areas that don't stand out, and not in beautiful glens or historical places, they are acceptable, but it seems to be a free-for-all, building them in some beautiful glens and historic sites.

I told my wife when my time is up to scatter my ashes at a place called Beaton's Bridge (not its proper name, but it is what I called it when I was growing up, as the farmer who worked the farm was called Beaton). It was where I poached my first salmon when I was very young. In my young days the river to me was the size of the Tay; in fact it is only about three metres wide at that point. We went there when we were taking photos for the book. To my horror, at the far side of the bridge was a turbine house for a hydro scheme. I will have to find a new place for my ashes (any suggestions?). To be fair this hydro scheme is hidden in the wood and is very well concealed.

Here I am now, sitting on my chair, with a nice wee dram, looking out at the beautiful view of the river and the mountains, with Alexa in the background saying, 'Peter, remember to take your tablets.'

61. Peter and Paulette
Reproduced by kind permission of Valentin Petrov (the artist)

ABOUT THE AUTHOR

Peter Weir was born at The Lodge, Castle Lachlan, Strathlachlan. He went to Strathlachlan Primary School, where there were about seven to ten pupils. He attended Dunoon Grammar School and left the school age fifteen (because he was fed up standing in the corner getting the belt).

Peter worked on the Castle Lachlan farm during his teen years. He then worked on road construction, as well as working on construction in London. Later, he worked in forestry as a ploughman, and in other forestry jobs. He was the manager of Glencoe Ski Centre for 11 years and served as a member of Glencoe Mountain Rescue Team for 21 years.

During the years Peter was on the mountain rescue team he worked on several climbing films, on BBC television programmes such as *Blue Peter* and BBC Television News, and also on the Hollywood movies *Highlander* and *Rob Roy*.

Peter built and ran a guest house for some years. In 2002, he helped to build a restaurant with rooms, and then learned to cook while working alongside his wife in the restaurant.

www.ingramcontent.com/pod-product-compliance
Lightning Source LLC
Chambersburg PA
CBHW071247070526
44583CB00017B/2364